The Essentials of Organic Chemistry

by

Adrian V. George
Leslie D. Field
Trevor W. Hambley

A supplementary text on fundamental
aspects of organic chemistry

© 1996 by Prentice Hall Australia Pty Ltd

All rights reserved. No part of this publication may be reproduced, stored in a retrieval system, or transmitted in any form or by any means, electronic, mechanical, photocopying, recording, or otherwise, without written permission of the publisher.

Acquisitions Editor: Kaylie Smith
Production Editor: Elizabeth Thomas

Typeset by DOCUPRO, Sydney
Printed in Australia by Star Printery Ptd Ltd, Erskineville, NSW

2 3 4 5 00

ISBN 0 7248 0906 6

National Library of Australia
Cataloguing-in-Publication Data

Field, L. D. (Leslie David)
  The essentials of organic chemistry

  Includes index
  ISBN 0 7248 0906 6

  1. Chemistry, Organic. I. Hambley, Trevor W. (Trevor William), 1955– . II. George, Adrian V. (Adrian Valentine), 1960– . III. Title.

547

Prentice Hall of Australia Pty Ltd, *Sydney*
Prentice Hall, Inc., *Eagelwood Cliffs, New Jersey*
Prentice Hall Canada, Inc., *Toronto*
Prentice Hall Hispanoamericana, *SA, Mexico*
Prentice Hall of India Private Ltd, *New Delhi*
Prentice Hall International, Inc, *London*
Prentice Hall of Japan, Inc., *Tokyo*
Prentice Hall of Southeast Asia Pty Ltd, *Singapore*
Editora Prentice Hall do Brasil Ltda, *Rio de Janeiro*

PRENTICE HALL

Preface

We have written *The Essentials of Organic Chemistry* as there is insufficient coverage of organic chemistry currently provided in all of the widely used first-year university general chemistry textbooks. This deficiency is usually covered by recommending a comprehensive organic text for first-year students, many of whom do not proceed to chemistry majors. *The Essentials of Organic Chemistry* is designed to supplement general chemistry texts and to provide a coverage of organic chemistry to meet the needs of most first-year university courses in chemistry. This book is not designed to stand alone as a chemistry text and relies on the coverage of such areas as bonding and hybridisation provided in most of the general chemistry texts. *The Essentials of Organic Chemistry* has been written in the style of *Chemistry, The Central Science* by Brown, LeMay and Bursten but might also be used with other widely used texts.

We wish to thank the staff at Prentice Hall Australia for their assistance and motivation. Without their support the project would not have proceeded. In particular we wish to acknowledge Andrew Binnie and David Weston who initiated the project and Kaylie Smith who shepherded it through to completion.

The authors invite comment from students and staff about this text.

Adrian George
Leslie Field
Trevor Hambley

Contents

1 Organic Hydrocarbons 1

1.1 Alkanes 3
    Structures of alkanes 5
    Constitutional isomers 10
    Naming alkanes 10
    Cycloalkanes 15
    Reactions of alkanes 16
    Functional groups 19

1.2 Alkenes and Alkynes 22
    Alkenes 22
    Alkynes 25
    Reactions of alkenes and alkynes 28
    Polymerisation of alkenes 34

1.3 Aromatic hydrocarbons 36
    Electrophilic substitution reactions of aromatic hydrocarbons 37
    Multiple substitutions and the directing influence of groups 42
    Oxidation of side chains in alkylbenzenes 43

Summary 46
Key terms 48
Exercises 48

2 Derivatives of Hydrocarbons 55

2.1 Halogen compounds 56
    Naming halogen compounds 56
    Bonding and properties of alkyl halogen compounds 56
    Nucleophilic substitution reactions 57
    Elimination reactions of alkyl halides 59
    Formation of Grignard reagents 62

2.2	**Alcohols**	**65**
	Naming alcohols	65
	Bonding and properties of alcohols	66
	Formation of alkoxides	67
	Oxidation of alcohols	68
	Conversion of alcohols to alkyl halides	71
	Dehydration of alcohols to alkenes	72
2.3	**Phenols**	**75**
2.4	**Ethers**	**75**
	Properties and reactions of ethers	78
2.5	**Amines**	**79**
	Naming amine compounds	80
	Properties and reactions of amines	81
	Basicity of amines	81
	Alkylation of amines	82
	Aromatic amines	86
	Formation of diazonium ions and reactions	86
2.6	**Aldehydes and ketones**	**89**
	Naming carbonyl compounds	89
	Bonding in aldehydes and ketones	91
	Nucleophilic addition reactions of aldehydes and ketones	91
	Oxidation of aldehydes	102
	Catalytic reduction of aldehydes and ketones	103
2.7	**Carboxylic acids and carboxylic acid derivatives**	**104**
	Naming carboxylic acids	104
	Properties and reactions of carboxylic acids	105
	Carboxylic acid derivatives	107
	Formation of carboxylic acids	112
	Reduction of carboxylic acids and esters to primary alcohol	113
Summary		**115**
Key terms		**117**
Exercises		**117**

3 Separation and identification of organic compounds — 125

3.1 Separation and purification of organic compounds — 126

3.2 Identification of organic compounds — 130
Elemental analysis — 130
Mass spectrometry — 131
Absorption spectroscopy — 135
Ultraviolet-visible spectroscopy — 136
Infrared spectroscopy — 139
Nuclear Magnetic Resonance (NMR) spectroscopy — 141
Crystal structure analysis — 147

3.3 The three-dimensional structure of organic compounds —stereochemistry of organic compounds — 148
Optical isomerism — 149
The polarimeter — 152
The absolute configuration at a stereogenic centre — 154
Molecules with more than one stereogenic centre — 159
Resolution of a mixture of enantiomers — 161

Summary — 161
Key terms — 162
Exercises — 163

Glossary — 172
Index — 178

Organic Hydrocarbons 1

Organic chemistry deals with compounds in which carbon is the principal element. Carbon forms a vast number of compounds; many millions of organic compounds are known, and about 90% of all new compounds prepared each year contain carbon. Consequently, the study of carbon compounds has come to constitute a separate branch of chemistry.

The terms *organic chemistry* and *organic compounds* are historical. They arose in the eighteenth century from the 'vitalist theory,' which held that organic compounds could be formed only by living organisms. In 1828, Friedrich Wöhler, a German chemist, reacted two non-organic compounds which had never been part of a living system, potassium cyanate (KOCN) and ammonium chloride (NH_4Cl), and obtained urea (H_2NCONH_2). Urea was a well-known organic substance that had been isolated from the urine of mammals. This was the first-documented instance in which an organic compound had been produced from inorganic materials. Today, although there is a vast number of organic compounds obtained from natural sources, the majority of organic compounds are synthetic and chemists prepare many organic compounds from inorganic and organic starting materials.

Because the compounds of carbon are so numerous, it is convenient to organise them into families that exhibit structural similarities. The simplest class of organic compounds is the **hydrocarbons**, those composed only of carbon and hydrogen. Organic compounds containing other elements can be considered to be derivatives of hydrocarbons.

The key structural feature of hydrocarbons, and most other organic substances, is the presence of stable carbon-carbon C—C bonds. The ability of carbon to form stable, extended chains of atoms bonded through single, double, or triple bonds is almost unique among the non-metallic elements. Hydrocarbons are divided into four classes on the basis of the kinds of C—C bonds in their molecules. Figure 1.1 shows an example of each of the four types of hydrocarbons: alkanes, alkenes, alkynes, and aromatic hydrocarbons.

Alkanes are hydrocarbons that contain only single bonds, as in ethane, C_2H_6. Because alkanes contain the largest possible number of atoms bonded to each carbon atom, they are called *saturated hydrocarbons*. Notice that each carbon atom in an alkane has four single bonds, whereas each hydrogen atom forms one single bond. **Alkenes**, also known as *olefins*, are hydrocarbons with one or more C—C double bonds. **Alkynes** contain at least one C—C triple bond. In **aromatic hydrocarbons**, the carbon atoms are connected in a planar ring structure, joined by a combination of double and single bonds between carbon atoms. Benzene, C_6H_6, is the best-known example of an aromatic hydrocarbon. Alkenes, alkynes, and aromatic hydrocarbons are called *unsaturated hydrocarbons* because they contain fewer attached atoms than an alkane having the same number of carbon atoms.

The members of the different series of hydrocarbons exhibit different chemical

Alkanes

methane	CH_4	CH_4
butane	$CH_3-CH_2-CH_2-CH_3$	C_4H_{10}

Alkenes

ethylene	$CH_2=CH_2$	C_2H_4
propene	$CH_3-CH=CH_2$	C_3H_6

Alkynes

acetylene	$H-C{\equiv}C-H$	C_2H_2
2-butyne	$CH_3-C{\equiv}C-CH_3$	C_4H_6

Aromatic hydrocarbons

benzene — C_6H_6

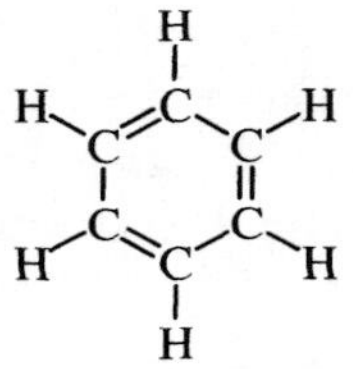

Figure 1.1 Names, geometrical structures, and molecular formulas for examples of alkanes, alkenes, alkynes, and aromatic hydrocarbons

properties; however, they are alike in many ways. Because carbon and hydrogen do not differ greatly in electronegativity (2.5 for carbon, 2.2 for hydrogen), the carbon—hydrogen bond is relatively non-polar and consequently hydrocarbon molecules are relatively non-polar. Hydrocarbons are almost insoluble in water, but they dissolve readily in non-polar solvents. Furthermore, hydrocarbons become less volatile with increasing molar mass because of an increase in the London dispersion forces.

1.1 ALKANES

Table 1.1 lists examples of the simplest alkanes. We are familiar with many of these substances because of their widespread use. Methane is the major component of natural gas which is used for cooking on gas stoves and for water heaters. Propane is the major component of bottled gas and is used as an automotive fuel (LPG), for heating and for cooking in areas where natural gas is not available. Butane is used in disposable lighters

Table 1.1 Several members of the series of straight-chain alkanes

Molecular formula	Condensed structural formula	Name	Boiling point (°C)
CH_4	CH_4	Methane	−161
C_2H_6	CH_3CH_3	Ethane	−89
C_3H_8	$CH_3CH_2CH_3$	Propane	−44
C_4H_{10}	$CH_3CH_2CH_2CH_3$	Butane	−0.5
C_5H_{12}	$CH_3CH_2CH_2CH_2CH_3$	Pentane	36
C_6H_{14}	$CH_3CH_2CH_2CH_2CH_2CH_3$	Hexane	68
C_7H_{16}	$CH_3CH_2CH_2CH_2CH_2CH_2CH_3$	Heptane	98
C_8H_{18}	$CH_3CH_2CH_2CH_2CH_2CH_2CH_2CH_3$	Octane	125
C_9H_{20}	$CH_3CH_2CH_2CH_2CH_2CH_2CH_2CH_2CH_3$	Nonane	151
$C_{10}H_{22}$	$CH_3CH_2CH_2CH_2CH_2CH_2CH_2CH_2CH_2CH_3$	Decane	174
$C_{15}H_{32}$	$CH_3(CH_2)_{13}CH_3$	Pentadecane	270
$C_{20}H_{42}$	$CH_3(CH_2)_{18}CH_3$	Eicosane	343

and in fuel canisters for camping stoves and lanterns which operate on gas. Alkanes with 5 to 12 carbon atoms per molecule are liquids and are commonly found in petrol (gasoline). Those with more than 20 carbon atoms are waxy solids.

The formulas for the alkanes given in Table 1.1 are written in a notation called a *condensed structural formula*. This notation reveals the way in which atoms are bonded to one another, but it does not require all of the bonds to be drawn. For example, the structural formula and condensed structural formula for butane, C_4H_{10}, are:

$$\begin{array}{c} \text{H} \quad \text{H} \quad \text{H} \quad \text{H} \\ | \quad | \quad | \quad | \\ \text{H}-\text{C}-\text{C}-\text{C}-\text{C}-\text{H} \\ | \quad | \quad | \quad | \\ \text{H} \quad \text{H} \quad \text{H} \quad \text{H} \end{array} \qquad CH_3CH_2CH_2CH_3$$

structural formula   condensed structural formula

Either the structural formulas or condensed structural formulas are common ways to represent organic compounds.

Each succeeding compound in the series listed in Table 1.1 has an additional CH_2 unit. A group of compounds, such as that shown in Table 1.1, is known as a **homologous series** because the same general formula can be used to describe all members of the group. The general formula for all the compounds listed in Table 1.1 is C_nH_{2n+2}, where n is the number of carbon atoms. Ethane, for example, contains two carbon atoms (n = 2) and six hydrogen atoms (2n + 2 = 2 × 2 + 2 = 6).

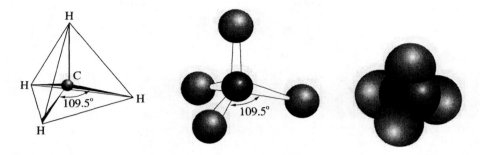

Figure 1.2 Representations of the three-dimensional arrangement of bonds about carbon in methane.

Structures of alkanes

The structural formulas and condensed structural formulas for alkanes do not indicate anything about the three-dimensional structures of these substances. The geometry about each carbon atom in an alkane is tetrahedral; that is, the four groups attached to each carbon are located at the corners of a tetrahedron. However, the three-dimensional structures are often represented as shown for methane in Figure 1.2. The bonding involves sp^3-hybridised orbitals on the carbon atom.

A CLOSER LOOK: Bonding in carbon compounds

An isolated, neutral carbon atom has four valence electrons and it needs to form four covalent bonds in order to achieve a closed shell (8-electron) configuration. These four bonds can be four single bonds, one double bond and two single bonds, one triple bond and one single bond, or two double bonds.

H–C(H)(H)–H	H₂C=O	H–C≡C–H	CH₂=C=CH₂
four single bonds	two single bonds one double bond	one single bond one triple bond	two double bonds

According to valence bond theory the hybridisation of the atomic orbitals of the carbon atoms in these four cases, will be sp^3, sp^2, sp, and sp respectively.

Valence shell electron pair repulsion theory (VSEPR) allows the prediction of the geometry that the hybridised orbitals will adopt. The basis of VSEPR theory is simply that the orbitals, since they are occupied by electrons, will arrange themselves as far away from each other as they can. If there are four orbitals, then they will adopt a tetrahedral geometry, three orbitals will adopt a trigonal planar geometry and two a linear geometry. A carbon atom that is sp^3 hybridised will have a tetrahedral geometry, one that is sp^2 hybridised will have trigonal planar geometry and one that is sp hybridised will have a linear geometry.

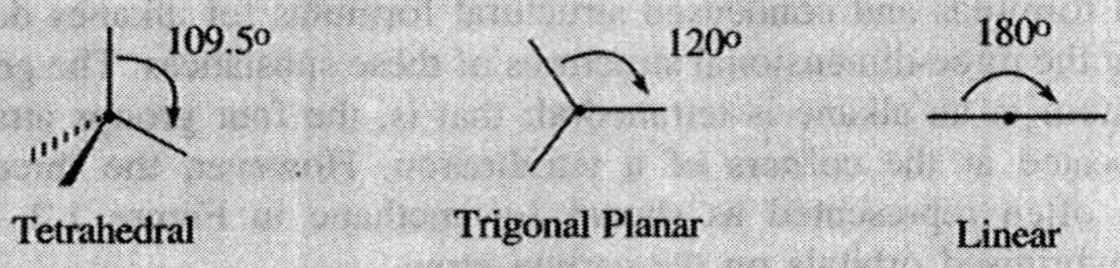

109.5°   120°   180°

Tetrahedral   Trigonal Planar   Linear

Rotation about the C—C single bond is a relatively facile process. You might imagine grasping the top left methyl group in Figure 1.3, which shows the structure of propane, and twisting it relative to the rest of the structure. Motion of this sort occurs very rapidly in alkanes at room temperature and a long-chain alkane is constantly undergoing motions that cause it to change its shape.

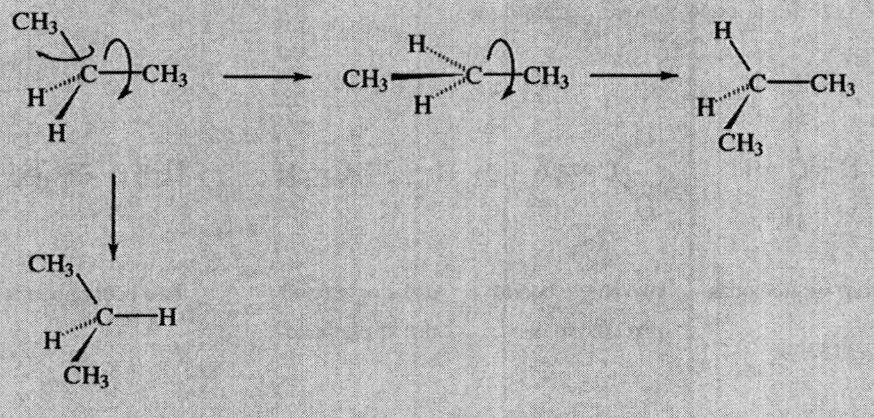

Figure 1.3 Three-dimensional models for propane, C_3H_8, showing rotations about the C—C single bonds

The rotation about any C—C single bond in an alkane can be examined in detail. When the carbon atom at one end of the C—C bond in ethane is rotated with respect to the other, the energy of the molecule changes as a function of the rotation angle because the hydrogen atoms occupy space and repel each other. The isomers generated simply by rotation about single bonds are called **conformational isomers**. If the carbon atoms of ethane are arranged so that the hydrogen atoms are as far apart as possible, then looking along the C—C bond the hydrogen atoms on one carbon atom appear to lie between those on the other carbon atom. This is the lowest energy arrangement for the ethane molecule and it is called the **staggered conformation**. If the carbon atoms are rotated with respect to each other such that the hydrogen atoms are as close as possible, this is the highest energy arrangement for the ethane molecule and it is called the **eclipsed conformation**.

The conformation of a molecule about any particular bond is best visualised as a *Newman projection*. This is a schematic view down the length of the bond, with the atoms at either end of the bond depicted as a point for the atom nearest the observer, and a circle for the atom away from the observer. Groups attached to the near atom radiate from the point and groups attached to the remote atom radiate from the periphery of the circle. Figure 1.4 shows Newman projections of the staggered and eclipsed conformations of ethane.

For the ethane molecule, the energy of the molecule is a function of the rotation angle with an energy maximum every 120° of rotation. The energy difference between the maxima and minima is small (only about 12 kJ mole^{-1}) and at room temperature the ends of the ethane molecule are rapidly rotating with respect to each other. If the hydrogen atoms are replaced by larger groups, the energy barrier can be significantly higher and the free rotation can be restricted or stopped. The larger the groups the larger the barrier, and the interaction of bulky groups with each other is called *steric hindrance*.

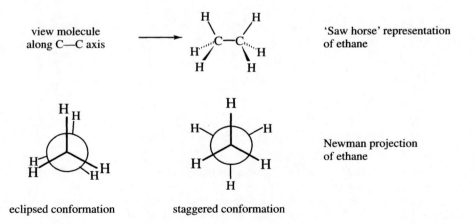

Figure 1.4 Newman projections of the eclipsed and staggered conformations of ethane

A CLOSER LOOK: The conformational isomers of butane

Conformational isomers (conformers) are those isomers that can be interconverted by rotation about single bonds. There is only a low energy barrier to rotation about a C—C single bond and the eclipsed and staggered conformers of ethane are known to rapidly interconvert in the gas phase. Each CH_3 group of the ethane molecule has three-fold symmetry so rotation about the C—C bond by 120° from one staggered conformation produces another (identical) staggered conformation.

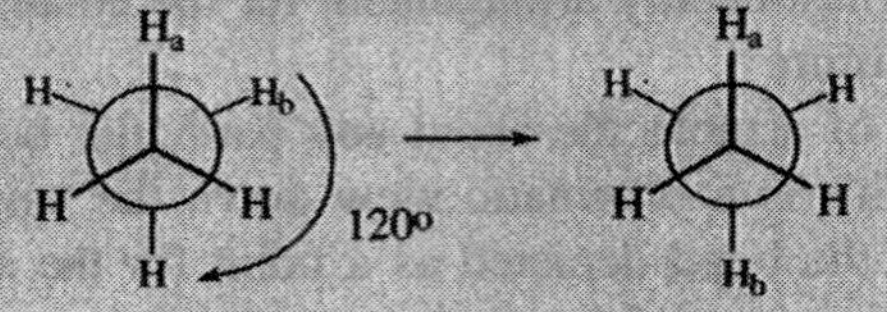

Butane can be considered as being derived from ethane by the replacement of one H atom on each C atom with a CH_3 group. This substitution destroys the three-fold symmetry that is present in ethane and, as a consequence, each rotation by 120° about the central C—C bond produces a different conformational isomer. The resulting conformations of butane are shown in the Newman projections below.

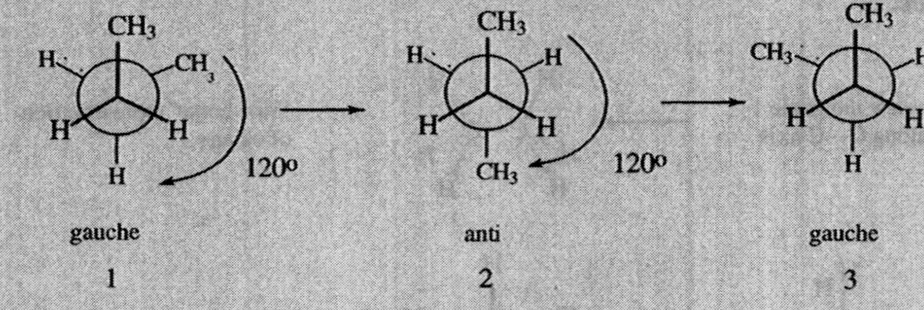

gauche   anti   gauche
1   2   3

The conformations 1 and 3 are labelled **gauche** and conformation 2 is labelled **anti**. The two gauche conformations, 1 and 3, are mirror images of each other and have the same energy. Conformation 2 (anti) has lower energy than the others because it has the relatively bulky CH_3 groups located as far apart from each other as is possible in the butane molecule.

There are also three eclipsed conformations:

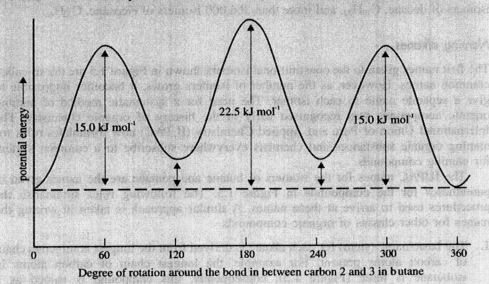

Conformations 4 and 6 are mirror images but in this case they have lower energy than conformation 5 because in 5 the bulky CH_3 groups are forced together. All of the eclipsed conformations have higher energy than the staggered conformations.

The energy of the butane molecules as a function of the 'torsion' angle about the central C—C bond is plotted below.

Constitutional isomers

The alkanes listed in Table 1.1 are termed *straight-chain* or *linear hydrocarbons* because all of the carbon atoms are joined in a single continuous chain. Alkanes consisting of four or more carbon atoms can also form branched chains; hydrocarbons with branched chains are called *branched hydrocarbons*. Figure 1.5 shows the condensed formulas and space-filling models for all the possible isomers of alkanes containing four and five carbon atoms. Notice that there are two ways in which four carbon atoms can be joined to give C_4H_{10}, either as a straight chain or a branched chain. For alkanes with five carbon atoms, C_5H_{12}, three different arrangements are possible.

Compounds with the same molecular formula, but with a different atom connectivity, are called **constitutional isomers**. The constitutional isomers of a given alkane have different physical properties to one another. Note that the melting point and boiling point of the isomers of butane, given in Figure 1.5, differ from those of the isomers of pentane. The number of possible isomers increases rapidly with the number of carbon atoms in the alkane. For example, there are 18 possible isomers of octane, C_8H_{18}, 75 possible isomers of decane, $C_{10}H_{22}$, and more than 366,000 isomers of eicosane, $C_{20}H_{42}$.

Naming alkanes

The first names given to the constitutional isomers shown in Figure 1.5 are the so-called common names. However, as the number of isomers grows, it becomes impossible to give a separate name to each isomer. The need for a systematic method of naming organic compounds was recognised early in the history of organic chemistry. The International Union of Pure and Applied Chemistry (IUPAC) now formulates rules for naming organic substances and chemists everywhere subscribe to a common system for naming compounds.

The IUPAC names for the isomers of butane and pentane are the names given in parentheses for the compounds in Figure 1.5. The following rules summarise the procedures used to arrive at these names. A similar approach is taken in writing the names for other classes of organic compounds.

1. The basic name (stem) for each alkane is derived from the longest continuous chain of carbon atoms present. For example, the longest chain of carbon atoms in isobutane is three (Figure 1.5); consequently, this compound is named as a derivative of propane, which has three carbon atoms.
2. In general, a group that is formed by removing a hydrogen atom from an alkane is called an **alkyl group**. The names for alkyl groups are derived by dropping the *-ane* ending from the name of the parent alkane and adding *-yl*. For example, the methyl group, CH_3, is derived from methane, CH_4. Likewise, the ethyl group, CH_3CH_2, is derived from ethane, C_2H_6. Table 1.2 lists the names and formulas of several of the more common alkyl groups. In the IUPAC system, isobutane is called methylpropane.
3. The location of an alkyl group along a chain of carbon atoms is indicated by numbering the carbon atoms along the chain. For example, the name 2-methylpentane indicates the presence of a methyl group, CH_3, on the second carbon atom of

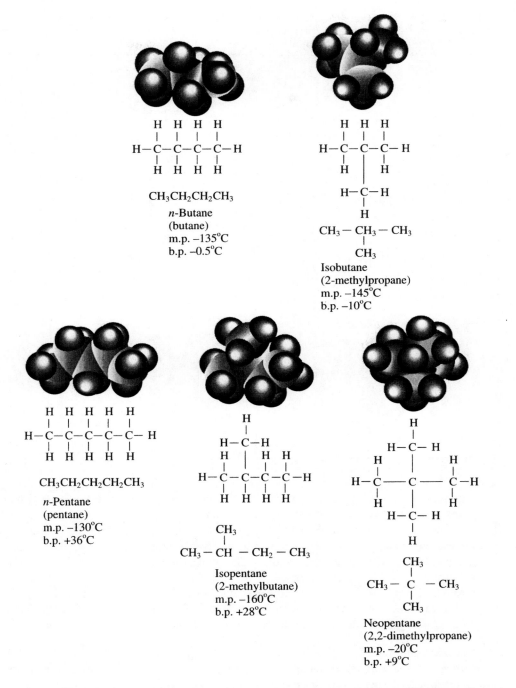

Figure 1.5 Space-filling models, names, structural formulas and melting/boiling points of the constitutional isomers of the alkanes C_4H_{10} and C_5H_{12}

Organic Hydrocarbons **11**

Table 1.2 Names and condensed structural formulas for several common alkyl groups

Group	Name
CH₃—	methyl
CH₃CH₂—	ethyl
CH₃CH₂CH₂—	propyl
CH₃CH₂CH₂CH₂—	butyl
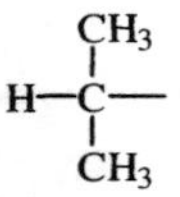 H—C(CH₃)(CH₃)—	isopropyl
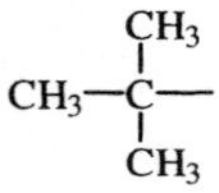 CH₃—C(CH₃)(CH₃)—	t-butyl

a pentane (five-carbon) chain. In general, the chain is numbered from the end that gives the lowest numbers for the alkyl positions.

4. If there is more than one substituent group of a certain type along the chain, the number of groups of that type are indicated by a prefix: *di-* (two), *tri-* (three), *tetra-* (four), *penta-* (five), and so forth. Therefore, the IUPAC name for neopentane (Figure 1.5) is 2,2-dimethylpropane. The prefix *di-* indicates the presence of two methyl groups; the 2,2- prefix indicates that both are on the second carbon atom of the propane chain. If two or more different substituent groups are present the names of the substituents are ordered alphabetically.

Organic compounds can also be drawn in an abridged shorthand form using *skeletal structures*. Skeletal structures enable the constitution of an organic compound to be conveyed quickly using a few simple rules:

1. The main chain of the structure is drawn as a zig-zag line. The segments of the

line represent C—C bonds. There is a carbon atom at each line junction and at the end of a line.
2. Hydrogen atoms are not shown, but the appropriate number of hydrogen atoms are *assumed* to be attached to each carbon atom to make up the valence of four.
3. Any non-carbon and non-hydrogen atoms are shown in full (for example, chlorine as Cl).

The skeletal structure conveys the same information as the condensed structure. It specifies what groups or atoms are present and the order in which they are connected to each other. The condensed and skeletal structures of 2-methylpentane are:

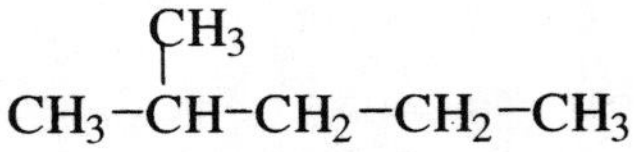

$$\begin{array}{c} CH_3 \\ | \\ CH_3-CH-CH_2-CH_2-CH_3 \end{array}$$

condensed structural formula 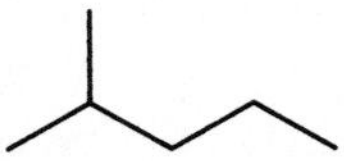   skeletal formula

SAMPLE EXERCISE 1.1

Name the following alkane:

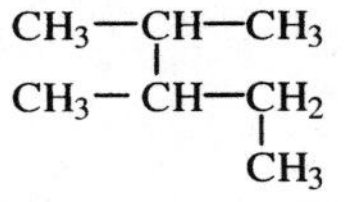

$$\begin{array}{c} CH_3-CH-CH_3 \\ | \\ CH_3-CH-CH_2 \\ | \\ CH_3 \end{array}$$

Solution: To name this compound properly, the longest continuous chain of carbon atoms must first be identified. This chain, extending from the upper left CH_3 group to the lower right CH_3 group, is five-carbon atoms long:

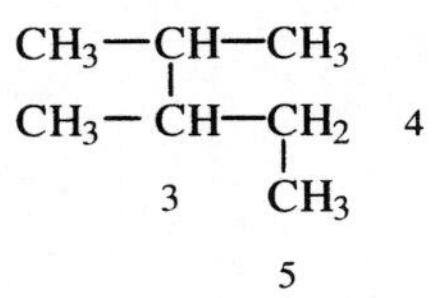

$$\begin{array}{c} \overset{1}{CH_3}-\overset{2}{CH}-CH_3 \\ | \\ CH_3-\underset{3}{CH}-\overset{4}{CH_2} \\ | \\ \underset{5}{CH_3} \end{array}$$

Therefore, the compound is named as a derivative of pentane. The carbon atoms could be numbered starting from either end. However, IUPAC rules state that the numbering should be chosen so that the numbers of those carbon atoms which bear side chains are as low as possible. The numbering should be started with the upper carbon in the diagram. There is a methyl group on carbon 2, and one on carbon 3. The compound is named 2,3-dimethylpentane. If the numbering were started from the lower carbon, the compound name would have been 3,4-dimethylpentane; this has higher numbers and is incorrect.

PRACTICE EXERCISE

Name the following alkane:

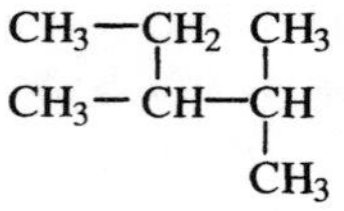

$$\begin{array}{c} CH_3-CH_2CH_3 \\ CH_3-CH-CH \\ CH_3 \end{array}$$

Answer: 2,3-dimethylpentane

SAMPLE EXERCISE 1.2

Write the condensed structural formula and skeletal formula for:

3-ethyl-2-methylpentane

Solution: The longest continuous chain of carbon atoms in this compound is five. To write the condensed formula, draw a string of five carbon atoms:

C—C—C—C—C

Next, place a methyl group on the second carbon atom, and an ethyl group on the middle carbon atom of the chain. Hydrogen atoms are then added to all the other carbon atoms to make up the four bonds to each carbon.

The condensed structural formula for 3-ethyl-2-methylpentane is:

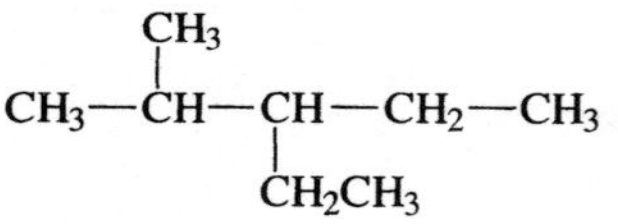

$$\begin{array}{c} CH_3 \\ CH_3-CH-CH-CH_2-CH_3 \qquad \text{or} \qquad CH_3CH(CH_3)CH(C_2H_5)CH_2CH_3 \\ CH_2CH_3 \end{array}$$

To draw the skeletal formula, first draw a zig-zag line of four segments (which implies carbon atoms at the three line junctions and the two ends of the chain). Attached at the first junction (the second carbon in the chain) a line is drawn (indicating a bond to one carbon atom). At the next junction (carbon 3) a zig-zag line with two segments is attached, indicating an ethyl group.

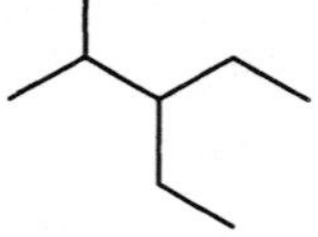

skeletal formula

PRACTICE EXERCISE

Write the condensed structural formula and the skeletal formula for 2,3-dimethylhexane.

Answer:

CH₃-CH-CH-CH₂-CH₂-CH₃ with CH₃ branch on C2 and CH₃ branch on C3

condensed structural formula   skeletal formula

Cycloalkanes

Alkanes can form not only branched chains, but rings or cycles as well. Alkanes with this form of structure are called **cycloalkanes**. Figure 1.6 illustrates a few examples of cycloalkanes. Cycloalkane structures are sometimes drawn as simple polygons in which each corner of the polygon represents a CH_2 group.

Carbon rings containing fewer than six carbon atoms are strained, because the C—C—C bond angle in the smaller rings must be less than the 109.5° tetrahedral angle that is ideal for tetravalent carbon atoms. The amount of strain is greatest for the smallest rings. In cyclopropane, which has the shape of an equilateral triangle, each C—C—C angle is only 60°; this molecule is much more reactive than either propane, its straight-chain analogue, or cyclohexane, which has no ring strain. The general formula for cycloalkanes is C_nH_{2n} (e.g. cyclohexane, C_6H_{12}). Hence, cycloalkanes form a separate homologous series.

Cycloalkanes are named according to the number of carbon atoms in the ring; if there are four carbon atoms, the ring is called a *cyclobutane*. If there is a single

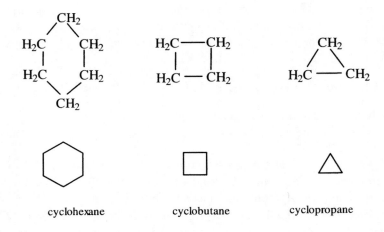

cyclohexane   cyclobutane   cyclopropane

Figure 1.6 Condensed structural formulas and skeletal formulas for three cycloalkanes.

substituent attached to the ring then its name is added as a prefix to the ring name. For example, the compound shown on the left in Figure 1.7 would be named *methylcyclobutane*. If there is more than one substituent attached to the ring then the carbon atoms of the ring are numbered.

Notice that although the two compounds on the right in Figure 1.7 are both called 1,2-dimethylcyclobutane, they differ in the arrangement of the methyl groups. These two compounds are examples of **stereoisomers**, compounds that have the same molecular formula and the same groups or atoms bonded to one another, but differ in the spatial arrangement of these groups or atoms. In the *cis* isomer, the two methyl groups are on the same face of the cyclobutane ring. In the *trans* isomer, the two methyl groups are on opposite sides. Stereoisomers possess distinct physical properties and can even differ significantly in their chemical behaviour.

Stereoisomerism in cycloalkanes arises because, unlike the C—C bond of a chain, the presence of the ring prevents rotation about the C—C bonds which make up the ring.

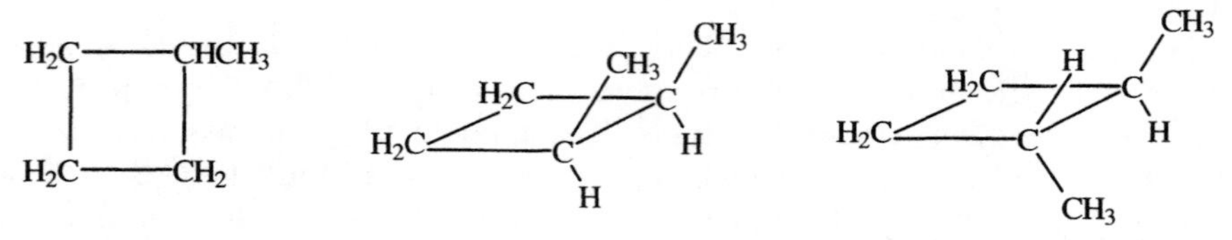

Figure 1.7 Condensed structural formulas for three substituted cyclobutanes.

Reactions of alkanes

Alkanes are relatively unreactive compounds and one reason for this is the stability of the C—C and C—H bonds. However, alkanes are not completely inert and one of the most important reactions of alkanes is **combustion** in air. For example, the complete combustion of methane proceeds as follows:

$$CH_4(g) + 2O_2(g) \rightarrow CO_2(g) + 2H_2O(g) \quad \Delta H^0 = -802 \text{ kJ mol}^{-1}$$

Methane is the major component of natural gas and the energy evolved in this reaction is widely used for heating. In this case we utilise the energy change rather than the products of the reaction.

Alkanes also undergo **substitution reactions** with F_2, Cl_2, and Br_2. In this reaction, halogen atoms replace one or more hydrogen atoms in the alkane. Reactions between F_2 and alkanes are very vigorous and special precautions must be taken to prevent explosions. Reactions with Cl_2 and Br_2 require heat or light to break Cl—Cl or Br—Br bonds in order to initiate the process. Light causes dissociation of the Cl_2 molecule, forming reactive Cl atoms:

$$Cl_2 \xrightarrow{\text{light}} 2Cl^{\bullet}$$

A chlorine atom has only seven valence-shell electrons. Species with an odd number of electrons are called **free radicals** (or simply radicals). They are frequently represented in chemical equations by showing a dot next to their chemical formula, representing the unpaired electron. Chlorine atoms are highly reactive and attack the alkane, removing a hydrogen atom and forming an alkyl radical. For example, when the alkane is ethane:

$$Cl^{\bullet} + CH_3CH_3 \longrightarrow HCl + CH_3CH_2^{\bullet}$$

The ethyl radical, $CH_3CH_2^{\bullet}$, combines with Cl_2 to give CH_3CH_2Cl and yet another chlorine atom:

$$CH_3CH_2^{\bullet} + Cl_2 \longrightarrow CH_3CH_2Cl + Cl^{\bullet}$$

The resultant chlorine atom reacts with more ethane, continuing the cycle. For each quantum of light absorbed by a chlorine molecule, many molecules of chloroethane can be formed. This reaction provides an example of a **radical chain** process. A disadvantage of radical chain reactions is that they are not highly selective. As the concentration of the chlorinated product builds up in the reaction, further chlorine atoms can abstract hydrogen atoms from the initial product, so that eventually dichloroethane and even more highly chlorinated molecules can be formed. A mixture of several products is usually formed in the reaction and these must be carefully separated when a single pure product is required.

Chemistry at Work

Petroleum, or crude oil, is a complex mixture of organic compounds, mainly hydrocarbons, with small quantities of other organic compounds containing nitrogen, oxygen, or sulfur. The tremendous demand for petroleum to meet the world's energy needs has led to the tapping of oil and gas wells in such forbidding places as the North Sea, northern Alaska, offshore in Bass Strait and off the coast of Western Australia.

The usual first step in the refining, or processing, of petroleum is to separate it into fractions on the basis of boiling point. The fractions commonly taken are shown in Table 1.3. Because petrol is the most commercially important of these fractions, petroleum refining processes are designed to maximise its yield.

Table 1.3 Hydrocarbon fractions from petroleum

Fraction	Size range of molecules	Boiling-point range (°C)	Uses
Gas	C_1 to C_5	−160 to 30	Gaseous fuel, production of H_2
Straight-run gasoline	C_5 to C_{12}	30 to 200	Petrol
Kerosene, fuel oil	C_{12} to C_{18}	180 to 400	Aviation fuel, diesel fuel, furnace fuel, cracking
Lubricants	C_{16} and up	350 and up	Lubricants and oils
Paraffins	C_{20} and up	Low melting solids	Candles
Asphalt	C_{36} and up	Gummy residues	Surfacing roads, fuel

Petrol is a mixture of volatile hydrocarbons containing varying amounts of aromatic hydrocarbons in addition to alkanes. In a car engine, a mixture of air and petrol vapour is compressed by a piston and then ignited by a spark plug. The burning of the alkane fuel should create a strong, smooth expansion of gas, forcing the piston outward and imparting force along the drive shaft of the engine. If the gas burns too rapidly, the piston receives a single hard slam rather than a strong, smooth push. The result is a 'knocking' or 'pinging' sound and a reduction in the efficiency with which energy produced by the combustion is converted to power.

The *octane number* of petrol is a measure of its resistance to knock. Petroleum with a higher octane number burns more smoothly and is a more effective fuel. The more highly branched alkanes have higher octane numbers than the straight-chain alkanes.

The octane number of petroleum is obtained by comparing its knocking characteristics to those of 'isooctane' (2,2,4-trimethylpentane) and heptane. Isooctane is assigned an octane number of 100, whereas heptane is assigned 0. Petroleum with the same knocking characteristics as a mixture of 90% isooctane and 10% heptane is rated as 90 octane.

The petrol obtained directly from fractionation of petroleum (called straight-run gasoline) contains mainly straight-chain hydrocarbons and has an octane number around 50. Straight run gasoline is subjected to a process called *cracking*, which converts the straight-chain alkanes into more desirable branched-chain alkanes.

Cracking is also used to convert some of the less volatile kerosene and fuel-oil fraction into compounds with lower molecular weights that are suitable for use as petrol. In the cracking process, the hydrocarbons are mixed with a catalyst and heated to 400–500°C. The catalysts used are naturally occurring clay minerals or synthetic Al_2O_3–SiO_2 mixtures. In addition for forming molecules suitable for petrol, cracking results in the formation of hydrocarbons of lower molecular weight, such as ethylene and propene. These substances are used industrially in a variety of processes to form plastics and other chemicals.

The octane rating of petroleum is further improved by adding certain compounds called *antiknock* agents. Until the mid-1980s, the principal antiknock agent was tetraethyllead, $(C_2H_5)_4Pb$. Its use has been drastically curtailed because of the environmental hazards of lead.

Functional groups

There are far too many organic compounds to study each one. Any organic compound can be viewed as a backbone (or skeleton) of carbon atoms bonded to each other with groups of atoms attached at various points. Each compound is unique, having its own properties (smell, boiling point, melting point, colour, and so on), but the compounds can be subdivided into classes on the basis of the nature of the groups of atoms attached to the skeleton. Although many groups are possible in theory, in practice there are a small number which are encountered in most organic compounds and these are called *functional groups*.

Functional groups confer characteristic chemical and physical properties on the compounds that contain them. Functional groups generally undergo the same chemical reactions irrespective of the type of molecule that contains them. The nature of the functional groups that an organic compound contains is the basis on which the compound is named.

Table 1.4 lists a number of the common functional groups which define classes of organic compounds. Any organic compound can contain one or more functional group. For example, all alcohols contain at least one —OH group and all ethers contain an oxygen atom directly bridging two carbon chains.

Table 1.4 Common functional groups which define classes of organic compounds

Class of organic compound	Functional group	R =	Example
alkene	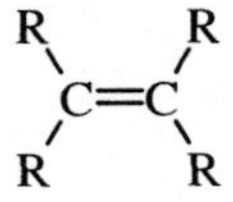 $R_2C=CR_2$	H, alkyl, aromatic	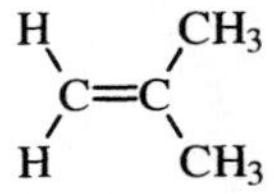 $H(CH_3)C=C(H)(CH_3)$
alkyne	$R-C\equiv C-R$	H, alkyl, aromatic	$H-C\equiv C-CH_2CH_3$
aromatic	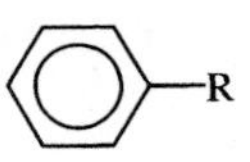 C_6H_5-R	H, alkyl, aromatic	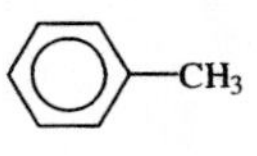 $C_6H_5-CH_3$
alkyl halide	$R-X$ X = F, Cl, Br, I	alkyl	CH_3CH_2-Br
alcohol	$R-OH$	alkyl	CH_3-OH

Table 1.4 *continued*

Class of organic compound	Functional group	R =	Example
ether	R—O—R	alkyl, aromatic	CH$_3$—O—CH$_3$
amine	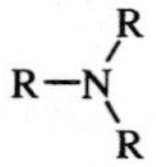 R—N(R)(R)	H, alkyl, aromatic	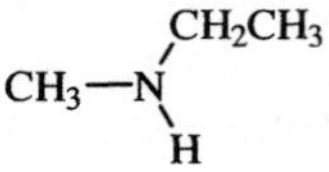 CH$_3$—N(CH$_2$CH$_3$)(H)
aldehyde	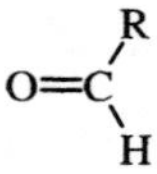 O=C(R)(H)	H, alkyl, aromatic	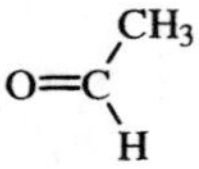 O=C(CH$_3$)(H)
ketone	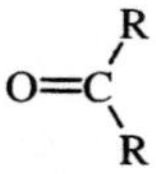 O=C(R)(R)	alkyl, aromatic	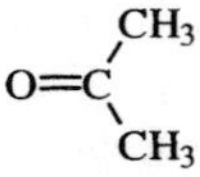 O=C(CH$_3$)(CH$_3$)
carboxylic acid	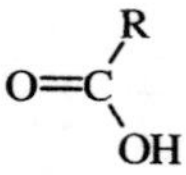 O=C(R)(OH)	H, alkyl, aromatic	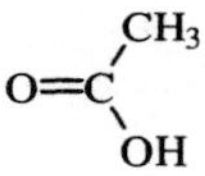 O=C(CH$_3$)(OH)
acid chloride	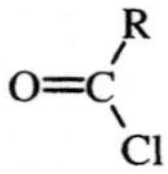 O=C(R)(Cl)	H, alkyl, aromatic	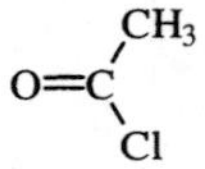 O=C(CH$_3$)(Cl)
acid anhydride	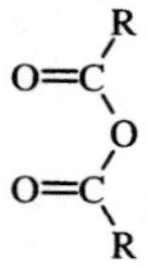 O=C(R)—O—C(=O)R	alkyl, aromatic	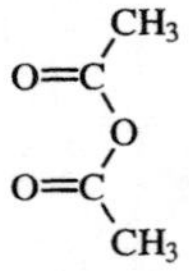 O=C(CH$_3$)—O—C(=O)CH$_3$
ester	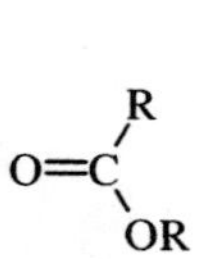 O=C(R)(OR)	alkyl, aromatic	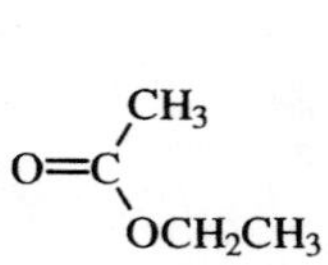 O=C(CH$_3$)(OCH$_2$CH$_3$)

Table 1.4 *continued*

Class of organic compound	Functional group	R =	Example
amide	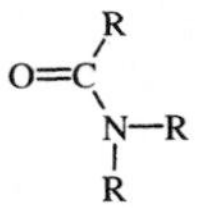 O=C(R)(N-R)(R)	H, alkyl, aromatic	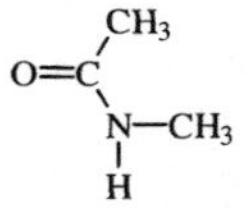 O=C(CH$_3$)(N-CH$_3$)(H)
nitrile	R—C≡N	alkyl, aromatic	CH$_3$—C≡N

Alkanes Around Us

2-Methyloctane: Many insects use chemicals to communicate with one another. Commonly, the substances are highly specific and operate at very low concentrations as aggregation agents, alarm pheromones, and population control substances. Mosquito larvae produce 2-methyloctane which inhibits the growth of younger larvae when there is a danger of overcrowding, thus ensuring sufficient food is available.

2-methyloctane

Steroids have a common tetracyclic structure composed of three cyclohexane rings (labelled A, B, and C) and a cyclopentane ring (labelled D) fused together. The group labelled R represents a wide selection of possible side chains. In humans most steroids function as chemical messengers, or hormones, and regulate many body processes. There are two main groups of steroid hormones: the sex hormones which regulate sexual development and reproduction, and the adrenocortical hormones which control metabolic processes.

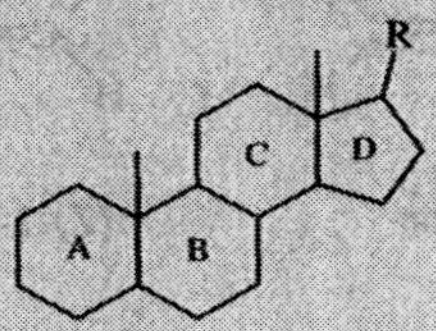

1.2 ALKENES AND ALKYNES

Alkenes

Alkenes are unsaturated hydrocarbons that contain one or more C—C double bonds (C=C). The presence of one or more multiple bonds makes unsaturated hydrocarbons significantly different from alkanes both in terms of their structure and their reactivity. The general formula for alkenes with one double bond is C_nH_{2n}. The simplest alkene, CH_2=CH_2, is called ethene or ethylene. The next member of the series, CH_3—CH=CH_2, is called propene or propylene. Alkenes with four or more carbon atoms have several isomers for each molecular formula. For example, there are four isomers of C_4H_8, as shown in Figure 1.8.

Alkene names are based on the longest continuous chain of carbon atoms that contains the double bond. The name given to this chain is obtained from the name of the corresponding alkane (Table 1.1) by changing the ending from -*ane* to -*ene*. The compound on the left in Figure 1.8 has a three-carbon chain containing the double bond, so the parent alkene is propene. It is not necessary to prefix methylpropene with 2 in this case because the point of attachment of the methyl group is unambiguous—attachment at either of the terminal carbon atoms would require the molecule to be named as a butene.

A number prefix locates the double bond along an alkene chain. The chain is always numbered from the end that gives the double bond the smallest number prefix. In propene, the only possible location for the double bond is between the first and second carbon atoms, so a prefix indicating its location is unnecessary.

If a substance contains two or more double bonds, each is located by a numerical prefix. Furthermore, the ending of the name is altered to identify the number of double bonds: -*diene* (two), -*triene* (three), and so forth. For example, CH_2=CH—CH_2—CH=CH_2 is 1,4-pentadiene.

Notice that the two compounds on the right in Figure 1.8 differ only in the relative locations of their terminal methyl groups. These two compounds are further examples of stereoisomers, compounds that have the same molecular formula and the same groups or atoms bonded to one another but differing in the spatial arrangement of these groups or atoms. In the *cis* or **Z** isomer, the two methyl groups are on the same side of the double bond, whereas in the *trans* or **E** isomer, the two methyl groups are on opposite sides. In formally naming alkenes, the prefixes Z or E are used to indicate clearly the stereochemistry of the double bond when there is the possibility of stereoisomerism.

methylpropene　　1-butene　　(Z)-2-butene　　(E)-2-butene

Figure 1.8 Condensed structural formulas and names of alkenes with molecular formula C_4H_8

A CLOSER LOOK: Bonding in alkenes and naming alkenes

Stereoisomerism in alkenes arises because, unlike the C—C single bond, the C=C double bond is resistant to twisting. The carbon atoms of the double bond of an alkene are sp^2-hybridised and the double bond of an alkene is comprised of a σ-component and a π-component. The π-component arises by overlap of *p*-orbitals of the carbon atoms. Rotation of the carbon atom at one end of a C=C with respect to the carbon at the other end requires that the π-bond is broken and this process involves considerable energy. For this reason, rotation about C—C double bonds does not occur under normal laboratory conditions.

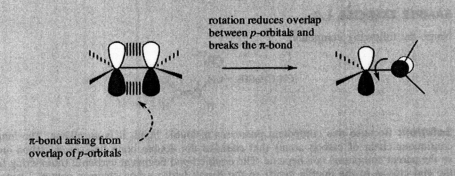

As a consequence of the restricted rotation, alkene compounds can form stereoisomers that can be separated and stored, and have different properties. The two stereoisomers in Figure 1.8 differ only in the relative positions of the methyl groups. However, these compounds are stable, isolable compounds with quite different physical properties.

The terms *cis* and *trans* had been used to differentiate the isomers; *cis* to indicate the isomer in which the substituents are on the same side of the double bond, *trans* when they are on opposite sides. However, the approved nomenclature for alkene stereoisomers now involves the prefixes Z or E. The carbon atoms at either end of the C=C are examined separately. Each of the carbon atoms of the C=C has two substituents and these are ranked according to a simple set of *priority rules* (or sequence rules). The priority of the two groups attached to a carbon atom of the C=C is based initially on the atomic number of the atom at the point of attachment. For example, a chlorine substituent would have a higher priority than a —CH$_3$ which in turn would have a higher priority than a —H. The sequence rules to cover the ranking of all substituents are treated in depth in Section 3.3. When the C=C is considered as a whole, if the groups with the highest priorities are on the same side of the double bond then the name of the alkene is prefixed with a Z (from the German *zusammen* meaning 'together'). If they are on opposite sides then the prefix is E (from *entgegen* meaning 'opposite'). If the two groups at one or other end of the double bond are identical (for example, both are H), then no stereoisomers arise from that double bond.

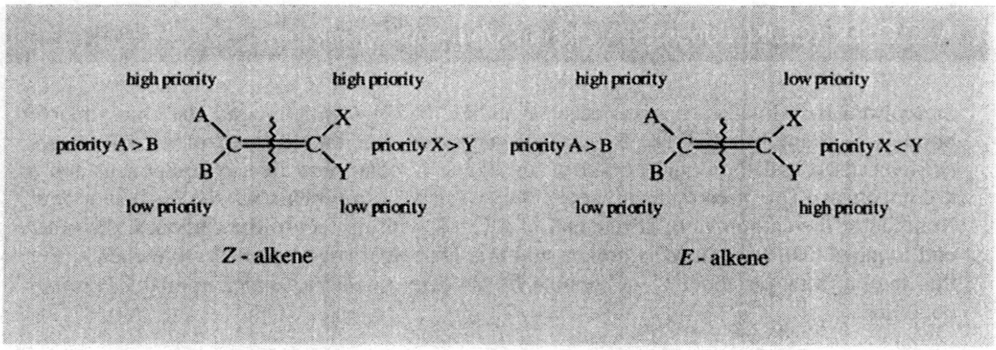

high priority, priority A > B, A, B low priority — C=C — high priority X, priority X > Y, Y low priority (Z-alkene)

high priority A, priority A > B, B low priority — C=C — low priority X, priority X < Y, Y high priority (E-alkene)

SAMPLE EXERCISE 1.3

Name the following compound:

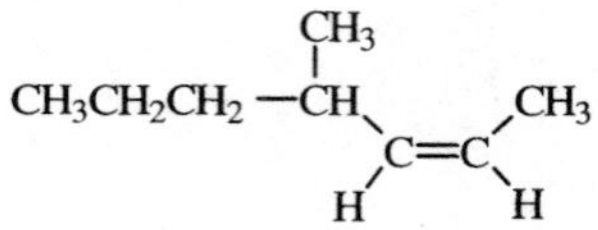

$$CH_3CH_2CH_2-CH(CH_3)-C(H)=C(H)-CH_3$$

Solution: Because this compound possesses a double bond, it is an alkene. The longest continuous chain of carbon atoms that contains the double bond has seven carbon atoms so the parent compound is a heptene. The double bond begins at carbon 2 (numbering from the end closest to the double bond) so the parent hydrocarbon chain is named 2-heptene. Continuing the numbering along the chain, a methyl group is bound at carbon atom 4 so the compound is 4-methyl-2-heptene. Finally, note that the geometrical arrangement of the alkyl groups bonded to the double bond places them on the same side of the C=C (in a *cis* arrangement). The complete name is *(Z)*-4-methyl-2-heptene.

PRACTICE EXERCISE

Draw the skeletal formula and the condensed structural formula for the compound *(E)*-5-methyl-2-hexene.

Answer:

(E)-5-methyl-2-hexene

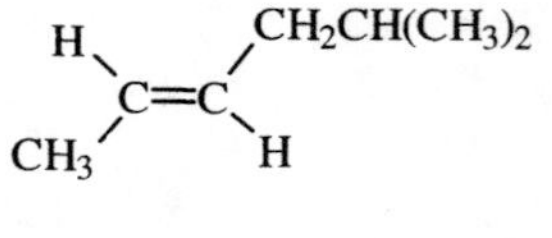

condensed structural formula

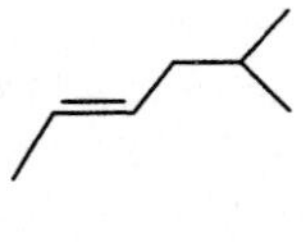

skeletal formula

Alkenes Around Us

Terpenes are natural products found in a wide variety of plants and animals. A characteristic feature of this class of molecules is that terpenes are formed from a combination of two or more C_5 isoprene units (isoprene is the common name for methyl-1,3-butadiene). Examples of terpenes include limonene which is the oil of lemon and orange, and α-pinene, one of the principal constituents of turpentine. The male cotton boll weevil (*Anthonomus grandis*) is attracted to the cotton plant by the monoterpenes such as limonene and α-pinene which the plant releases. Having found a suitable breeding site, the weevil releases aggregation pheromones of its own that attract both males and females of the same species. These pheromones are probably metabolic products derived from the terpenes in the plant, ingested during the initial feeding period.

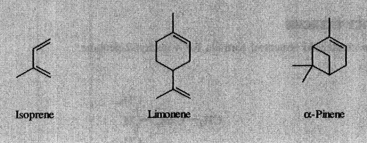

Isoprene   Limonene   α-Pinene

Alkynes

Alkynes are unsaturated hydrocarbons containing one or more C≡C bonds. The general formula for straight-chain alkynes containing one triple bond is C_nH_{2n-2}. The simplest alkyne is acetylene, C_2H_2, a highly reactive molecule. When acetylene is burned in a stream of oxygen, for example, in an oxyacetylene torch, the flame reaches a temperature of about 3200K. Alkynes in general are highly reactive molecules and, because of this, they are important intermediates in many industrial processes.

Alkynes are named by identifying the longest continuous chain in the molecule containing the triple bond and modifying the ending of the name as listed in Table 1.1 from *-ane* to *-yne*.

SAMPLE EXERCISE 1.4

Name the following compounds:

(a) $CH_3CH_2CH_2-C\equiv C-CH_3$

(b)   CH₃CHCH₂CH₂—C≡CH
         |
         CH₂CH₃

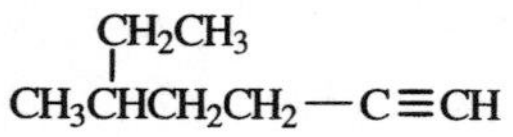

Solution: In (a) the longest chain contains six carbon atoms. There are no side chains. The triple bond begins at carbon 2 (remember that the numbering is arranged so that the smallest possible number is assigned to the carbon containing the multiple bond). The complete name is 2-hexyne.

In (b) the longest continuous carbon chain which contains the triple bond has seven carbon atoms so this compound is named as a derivative of heptyne. The complete name is 5-methyl-1-heptyne.

PRACTICE EXERCISE

Give the condensed structural formula for 4-methyl-2-pentyne.

Answer:

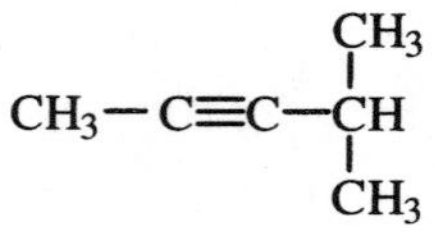

$$CH_3-C\equiv C-\underset{\underset{CH_3}{|}}{\overset{\overset{CH_3}{|}}{CH}}$$

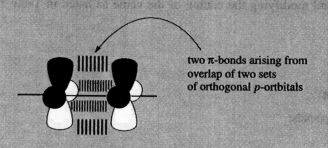

A CLOSER LOOK: Bonding in alkynes

A triple bond is composed of one σ-bond and two π-bonds. The two carbon atoms of the C≡C triple bond are *sp*-hybridised and each possess two *p*-orbitals which combine to form two π-bonds. The distribution of atoms around each of the carbon atoms of the triple bond is linear and consequently alkynes do not show stereoisomerism.

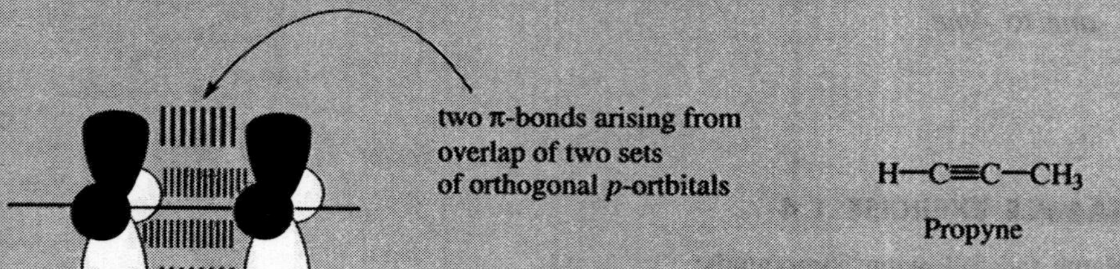

two π-bonds arising from overlap of two sets of orthogonal *p*-orbitals

H—C≡C—CH₃

Propyne

no sterioisomers possible

26   The Essentials of Organic Chemistry

Alkynes Around Us

Dehydromatricaria ester (methyl dec-2-en-4,6,8-triynoate): This compound is isolated from daisy seeds. The alkyne functional group is particularly common in the plant families *Compositae* (daisy) and *Umbelliferae* (carrot and parsley). Although the occurrence of alkynes in nature is widespread, little is known about the action of many of the compounds. The naturally occurring alkynes are frequently light and air sensitive. Some show potent antibiotic activity, but their toxicity precludes them from therapeutic use.

$$CH_3-C\equiv C-C\equiv C-C\equiv C-CH\overset{E}{=}CH-COOCH_3$$

dehydromatricaria ester

Dynemicin is one of a series of molecules which can invade and destroy cancer cells. Once in the host cell, the *enediyne* fragment rearranges to form a benzene ring containing two free radicals. These free radicals scavenge hydrogen atoms from the DNA of the cell, breaking the DNA and destroying the cell in the process. Although the enediyne is the active centre of dynemicin, the other groups in the molecule are vitally important to convey the molecule to a specific target—the cancer cell, tether the group in place and trigger the enediyne to rearrange to the benzene diradical.

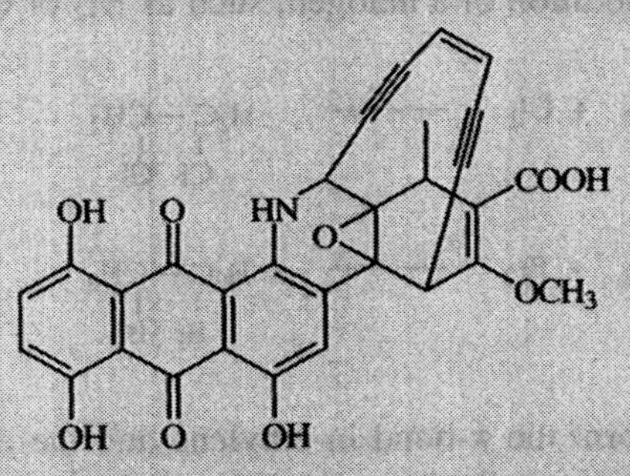

Dynemicin A

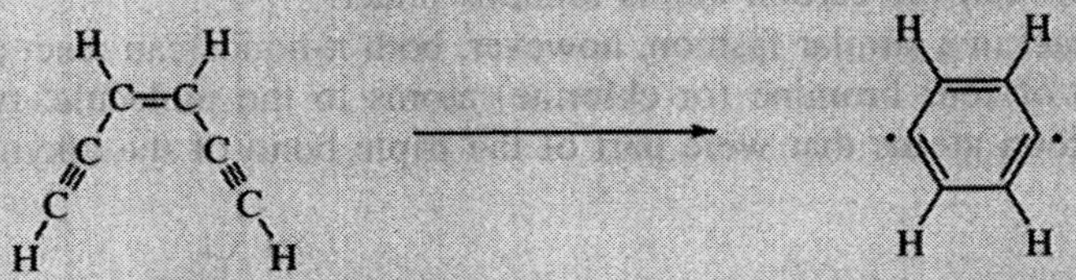

an enediyne → benzene diradical

Alkynes with the C≡C at the end of a carbon chain, that is, containing the —C≡C—H group are called *terminal alkynes*. The hydrogen attached to the *sp*-hybridised carbon atom is more acidic than in other hydrocarbons and can be removed by treatment with a very strong base. Hydroxide is about 10^7 times too weak to remove the terminal hydrogen as H⁺ from a terminal alkyne; however, bases such as sodium amide or butyllithium, will readily deprotonate terminal alkynes to form alkynide ions. Alkynide ions formed from terminal alkynes are useful intermediates in the synthesis of other alkynes (see Section 2.1).

$$R-C\equiv C-H \xrightarrow[e.g.\ NaNH_2\ in\ liquid\ NH_3]{strong\ base} R-C\equiv C^-$$

a terminal alkyne → an alkynide ion

$$CH_3-C\equiv C-H \xrightarrow{NaNH_2\ in\ liquid\ NH_3} CH_3-C\equiv C^-$$

propyne → propynide ion

Reactions of alkenes and alkynes

The presence of C—C double or triple bonds in hydrocarbons makes them more reactive than alkanes. The most characteristic reactions of alkenes and alkynes are **addition reactions**, in which a reactant is added to the two carbon atoms that form the multiple bond. A simple example is the addition of a halogen, such as Br_2 or Cl_2, to ethylene:

$$H_2C=CH_2 + Cl_2 \longrightarrow \underset{Cl\ \ Cl}{H_2C-CH_2}$$

$$H_2C=CH_2 + Br_2 \longrightarrow \underset{Br\ \ Br}{H_2C-CH_2}$$

The pair of electrons that form the π-bond in ethylene and the electron pair in the σ-bond of bromine are used in forming the two new σ-bonds to the two bromine atoms. The σ-bond between the carbon atoms remains intact.

Alkynes react in a similar fashion, however, both π-bonds can react and this results in the addition of four bromine (or chlorine) atoms to the molecule, two attached to each of the carbon atoms that were part of the triple bond of the alkyne.

$$CH_3-C\equiv C-H + 2Cl_2 \longrightarrow CH_3-\underset{Cl\ \ \ Cl}{\overset{Cl\ \ \ Cl}{C-C}}-H$$

The addition of H_2 to an alkene converts it to an alkane. This reaction, termed *hydrogenation*, does not occur readily at ordinary temperature and pressure. One reason for the lack of reactivity of H_2 towards alkenes is the high stability of the H—H bond. To promote the hydrogenation reaction, it is necessary to use a catalyst that assists in rupturing the H—H bond. The most widely used catalysts are finely divided metals (platinum, palladium, nickel, or palladium supported on carbon) onto which H_2 is adsorbed before it reacts with the double bond.

$$CH_3CH=CHCH_3 + H_2 \xrightarrow[\text{catalyst}]{\text{Pd/C}} CH_3CH_2CH_2CH_3$$

alkene → alkane

Alkynes are also hydrogenated completely with H_2 using a platinum or palladium catalyst to give alkanes. However, using a catalyst which has been partially deactivated (the addition of barium sulfate and quinoline deactivates a palladium catalyst), alkynes can be partially hydrogenated to alkenes. When non-terminal alkynes are hydrogenated to alkenes, the alkenes formed always have the Z-stereochemistry with both hydrogen atoms from H_2 attached to the same side of the double bond. When terminal alkynes are hydrogenated the alkene formed has two hydrogen atoms at one end of the double bond and hence no stereoisomers are possible.

$$CH_3CH_2-C{\equiv}C-H + H_2 \xrightarrow[\text{catalyst}]{\text{Pd}} CH_3CH_2CH_2CH_3$$

alkyne → alkane

$$CH_3-C{\equiv}C-CH_3 + H_2 \xrightarrow[\text{catalyst}]{\text{Pd/BaSO}_4 \text{ deactivated}} \begin{array}{c} CH_3 \quad\quad CH_3 \\ \diagdown\;\;\;\diagup \\ C{=}C \\ \diagup\;\;\;\diagdown \\ H \quad\quad H \end{array}$$

2-Butyne → (Z)-2-Butene

The addition of hydrogen halides (HCl, HBr, HI) and other non-symmetrical molecules to alkenes is more complicated than hydrogenation because two products are often possible. For example, addition of HBr to propene might produce either 1-bromopropane or 2-bromopropane:

$CH_3CH=CH_2$ + HBr
propene

2-bromopropane:
$$CH_3-\underset{Br}{\underset{|}{\overset{H}{\overset{|}{C}}}}-\underset{H}{\underset{|}{\overset{H}{\overset{|}{C}}}}-H$$

1-bromopropane:
$$CH_3-\underset{H}{\underset{|}{\overset{H}{\overset{|}{C}}}}-\underset{Br}{\underset{|}{\overset{H}{\overset{|}{C}}}}-H$$

In practice, only 2-bromopropane forms. In a large number of reactions where a hydrogen halide adds to an alkene it has been observed that only one of the two possible products is formed. This observation, known as **Markovnikov's rule**, was first formulated by the Russian chemist V. V. Markovnikov (1838–1904): When a hydrogen halide is added to an alkene, the hydrogen atom is usually attached to the carbon atom of the C=C that already has the most hydrogen atoms.

Markovnikov's rule enables us to predict which of the two possible products will predominate from the addition of a hydrogen halide to an alkene. The reason that Markovnikov's rule works is that the reaction proceeds in two distinct steps and the stability of the intermediate species formed in the reaction determines the products of the reaction. A **reaction mechanism** is determined from a careful study of the individual steps that lead from starting materials to products.

When an alkene reacts with a strong acid such as aqueous HCl, H$^+$ initially reacts with the C=C of the double bond to give a charged species called a **carbocation**. The pair of electrons in the π-bond of the C=C are used to form a new σ-bond to H$^+$. The H can be bonded to either of the carbon atoms which were part of the double bond, resulting in two possible products. Since both electrons of the π-bond are required to make the new bond to H, the carbon atom which does not become bonded to H, must have a positive charge. Figure 1.9 shows the two possible carbocations that could arise by reaction of propene with a strong acid.

The greater the number of alkyl groups (carbon chains) that are attached to the carbon which bears the positive charge, the more stable is the carbocation. This means that carbocations which have three alkyl groups attached to the carbon with the positive charge (called **tertiary carbocations** or 3° carbocations) are more stable than carbocations which have two alkyl groups attached to the positively charged centre (called **secondary carbocations** or 2° carbocations), and these in turn are more stable than

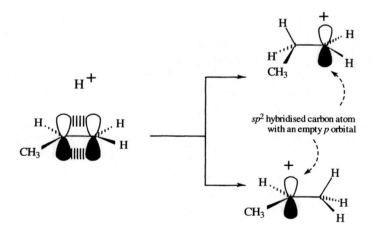

Figure 1.9 Two possible carbocations formed by reaction of propene with H$^+$.

those where only one alkyl group is attached to the positively charged centre (called **primary carbocations** or 1° carbocations).

$$\underset{\substack{3° \\ \text{carbocation}}}{\overset{R}{\underset{R}{\mid}}\underset{R}{\overset{\mid}{C^+}}} \quad \text{more stable than} \quad \underset{\substack{2° \\ \text{carbocation}}}{\overset{R}{\underset{R}{\mid}}\underset{H}{\overset{\mid}{C^+}}} \quad \text{more stable than} \quad \underset{\substack{1° \\ \text{carbocation}}}{\overset{R}{\underset{H}{\mid}}\underset{H}{\overset{\mid}{C^+}}}$$

If it is possible to form two different carbocations by the reaction of an alkene with a strong acid, the more stable carbocation of the two is preferentially formed (3° > 2° > 1°). Once formed, that carbocation reacts with any species in the reaction mixture which can attack the positively charged centre. Species which donate electron pairs and bond positively charged centres are termed **nucleophiles**. Species which accept electron pairs and bond to centres of high electron density (for example, to anions or to a double bond) are called **electrophiles**. If the strong acid used were HCl, then H⁺ would be an electrophile and Cl⁻ would be a nucleophile in the solution. So the reaction of HCl with 2-methylpropene would be written in detail:

$$\underset{CH_3}{\overset{CH_3}{\diagdown}}C=CH_2 \xrightarrow[\text{protonation with H}^+]{H^+} \underset{\substack{\text{more stable carbocation} \\ \text{intermediate}}}{\underset{CH_3}{\overset{CH_3}{\mid}}\underset{CH_3}{\overset{\mid}{C^+}}} \xrightarrow[\text{reaction with a nucleophile}]{Cl^-} \underset{Cl}{\overset{CH_3}{\underset{\mid}{CH_3-\overset{\mid}{C}-CH_3}}}$$

With an acid, such as aqueous H_2SO_4 as the catalyst, it is also possible to add H_2O to a double bond. The H_2SO_4 provides the source of H⁺ required for the initial protonation of the double bond to form a carbocation. The carbocation can be attacked by the oxygen atom of a water molecule because it has two lone pairs of electrons which makes it nucleophilic. The overall reaction sequence leads to the apparent addition of H—OH across the C=C bond of the alkene in the direction which would be predicted by Markovnikov's rule, that is, a hydrogen atom adds to the carbon with the most hydrogens, and the OH group adds to the carbon with the fewest hydrogens attached:

$$CH_3-CH=CH_2 \xrightarrow{H_2SO_4} \underset{\substack{\text{more stable} \\ \text{carbocation} \\ \text{intermediate}}}{CH_3-\overset{+}{C}H-CH_3} \xrightarrow[2. -H^+]{1. H_2O} \underset{}{\overset{OH}{\underset{}{CH_3-\overset{\mid}{C}H-CH_3}}}$$

Organic Hydrocarbons **31**

The addition of hydrogen halides to alkynes, resembles that of alkenes. Both π-bonds react and this results in the addition of two molecules of HX to each alkyne molecule. Markovnikov's rule applies to the reaction of terminal alkynes in the same way as it does to reactions of unsymmetrical alkenes.

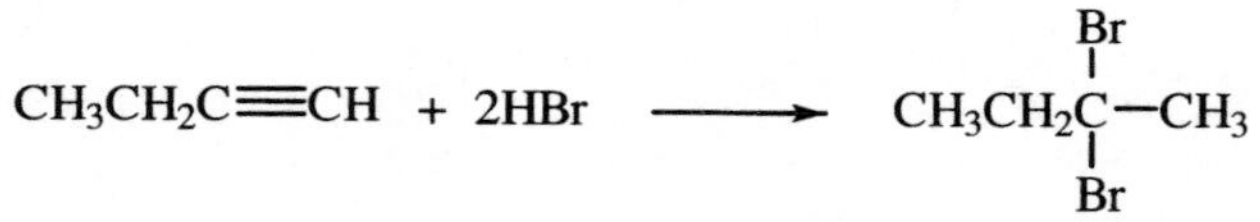

$$CH_3CH_2C{\equiv}CH + 2HBr \longrightarrow CH_3CH_2\underset{Br}{\overset{Br}{\underset{|}{\overset{|}{C}}}}-CH_3$$

SAMPLE EXERCISE 1.5

Predict the product of the reaction of HCl with the following compound:

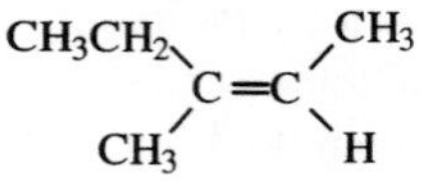

$$\underset{CH_3}{\overset{CH_3CH_2}{>}}C{=}C\underset{H}{\overset{CH_3}{<}}$$

Solution: One carbon atom of the C=C has a —H and a —CH$_3$ attached and the other has a —CH$_3$ group and a —CH$_2$CH$_3$ group. According to Markovnikov's rule, H⁺ will add to the carbon atom bearing the —CH$_3$, and —H and the chloride (derived from HCl) will add to the other carbon atom of the C=C. The major product will be 3-chloro-3-methylpentane:

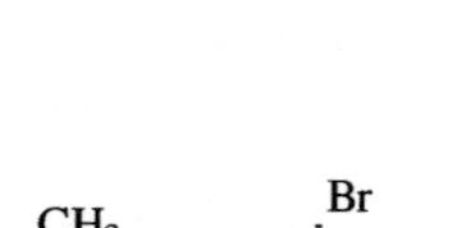

$$CH_3CH_2-\underset{CH_3}{\overset{Cl}{\underset{|}{\overset{|}{C}}}}-CH_2CH_3$$

PRACTICE EXERCISE

The compound:

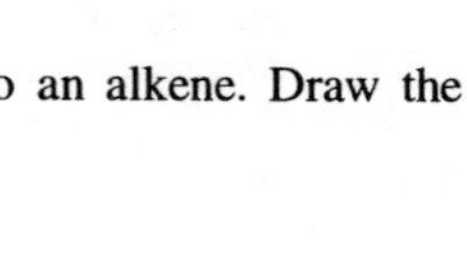

$$\underset{CH_3}{\overset{CH_3}{>}}\overset{}{\underset{CH_3}{\cdots}}C-C\underset{H}{\overset{Br}{<}}CH_3$$

was formed by addition of HBr to an alkene. Draw the condensed structural formula for the alkene.

Answer:

$$CH_3-\underset{\underset{CH_3}{|}}{\overset{\overset{CH_3}{|}}{C}}-CH=CH_2$$

Oxidative cleavage of alkenes can be achieved by the reaction with ozone (O_3). Ozone is prepared in the laboratory by passing a stream of oxygen through a high voltage electrical discharge, but it occurs naturally as a gas in the upper atmosphere. Ozone reacts rapidly with the C=C bond of alkenes to form an ozonide in which the carbon atoms that were part of the C=C are separated by —O— and —O—O— bridges. The ozonides can be explosive materials and are not usually isolated or stored, but instead they are treated with a mild reducing agent (such as a mixture of zinc and dilute acetic acid) to produce a mixture of aldehydes and/or ketones:

$$\underset{\text{an alkene}}{\overset{}{\text{C=C}}} \xrightarrow{O_3} \underset{\text{an ozonide}}{\overset{}{\text{C(O-O)C-O}}} \xrightarrow[\text{acetic acid}]{Zn} \underset{\text{aldehydes and/or ketones}}{\text{C=O + O=C}}$$

The overall reaction sequence, that is, addition of ozone then reduction, is called **ozonolysis**. The ozonolysis sequence can be used to locate the position of a double bond (or double bonds) in an alkene or diene chain because the aldehyde or ketone fragments that are formed can be isolated and identified, and the information pieced together to give the structure of the molecule which gave rise to them.

SAMPLE EXERCISE 1.6

Give the condensed structural formulas of the compounds formed when 2-methyl-2-butene is treated with ozone then with a mixture of zinc and acetic acid.

Solution: Ozone reacts at the C=C to form an ozonide which is reduced by the mixture of zinc and acetic acid. The C=C is cleaved and the carbon atoms which were originally part of the C=C are oxidised to an aldehyde and a ketone:

$$\underset{CH_3}{\overset{CH_3}{\text{C}}}=\underset{H}{\overset{CH_3}{\text{C}}} \xrightarrow[\text{2. Zn / acetic acid}]{\text{1. } O_3} \underset{CH_3}{\overset{CH_3}{\text{C}}}=O + O=\underset{H}{\overset{CH_3}{\text{C}}}$$

Organic Hydrocarbons

PRACTICE EXERCISE

Give the condensed structural formulas of the compounds formed when the following alkenes are treated with ozone then with a mixture of zinc and acetic acid:

(a) 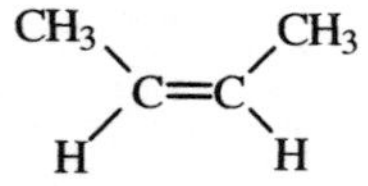 (CH₃)(H)C=C(CH₃)(H)   (b) (CH₃)(H)C=C(H)(CH₃)   (c) cyclohexylidene=CH₂

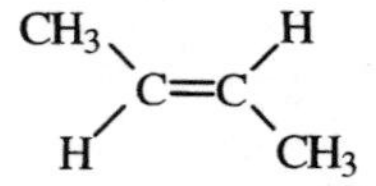

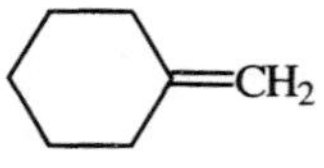

Answers:

(a) 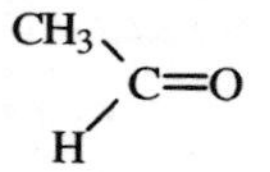 (CH₃)(H)C=O (2 moles)   (b) 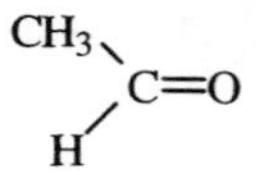 (CH₃)(H)C=O (2 moles)   (c) 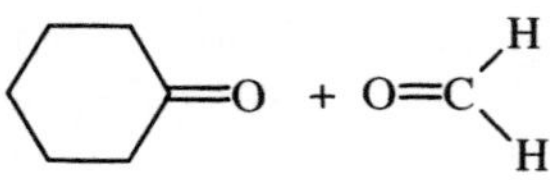 cyclohexanone =O + O=C(H)(H)

Polymerisation of alkenes

A **polymer** is a material with a high molecular weight, that is formed from simple molecules called **monomers**. Polymers may be either synthetic or natural in origin. Natural polymers include natural rubber, proteins, starch, and cellulose. Synthetic polymers include familiar materials such as Teflon and polystyrene.

Synthetic polymers are made by taking advantage of the chemical behaviour of the functional groups of their monomers. For example, monomers with C=C double bonds, such as tetrafluoroethylene, $F_2C=CF_2$, can be caused to react with one another by a series of addition reactions. The polymerisation of tetrafluoroethylene produces Teflon:

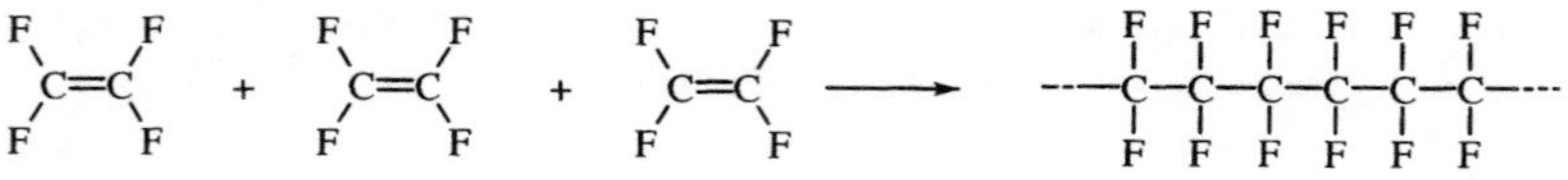

$F_2C=CF_2$ + $F_2C=CF_2$ + $F_2C=CF_2$ ⟶ ---CF₂-CF₂-CF₂-CF₂-CF₂-CF₂---

tetrafluoroethylene monomers   teflon polymer

In an *addition polymerisation* reaction, π-bonds are broken and the electrons used to form new σ-bonds between the monomer units. Many different alkenes have been used to synthesise polymers, as summarised in Table 1.5.

Table 1.5 Common addition polymers and their monomers

Structure of monomer	Structure of repeating unit	Name of polymer	Uses
$H_2C=CH_2$	$-(CH_2-CH_2)_n-$	Polyethylene	plastic packaging, bottles, cable insulation, plastic sheeting, unbreakable containers
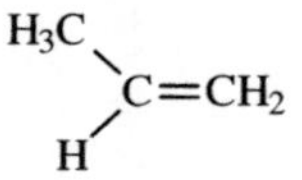 $H_3C\,\,\,\,\,\,\,\,\,\,\,\,\,\,\,\,\,$C=CH$_2$ / H	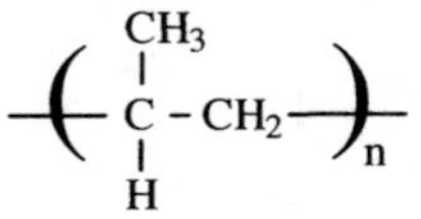 $-(\underset{H}{\overset{CH_3}{C}}-CH_2)_n-$	Polypropylene (Herculon)	automotive mouldings, carpet fibres, rope
$F_2C=CF_2$	$-(CF_2-CF_2)_n-$	Polytetrafluoroethylene (PTFE, teflon)	valves, gaskets, non-stick heat-resistant coatings
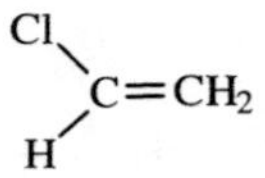 Cl\C=CH$_2$ / H	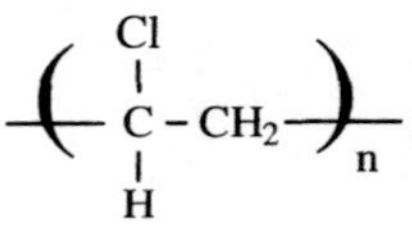 $-(\underset{H}{\overset{Cl}{C}}-CH_2)_n-$	Poly(vinyl chloride) (PVC)	pipe fittings, records, clear wrapping for meat products
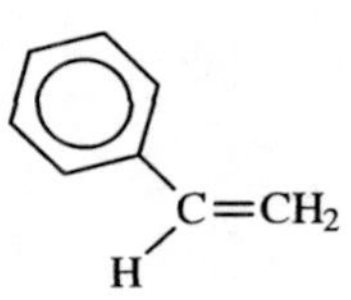 Ph\C=CH$_2$ / H	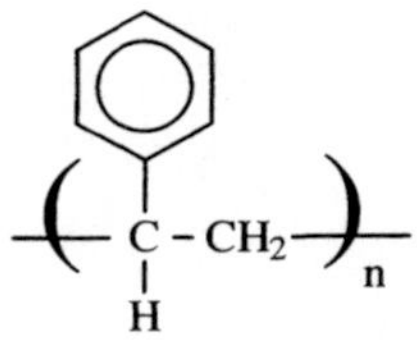 $-(\underset{H}{\overset{Ph}{C}}-CH_2)_n-$	Polystyrene (Styrofoam)	disposable food containers, insulation, packaging
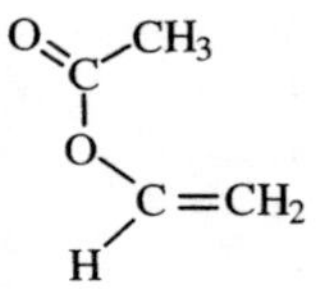 $O=C(CH_3)-O$\C=CH$_2$ / H	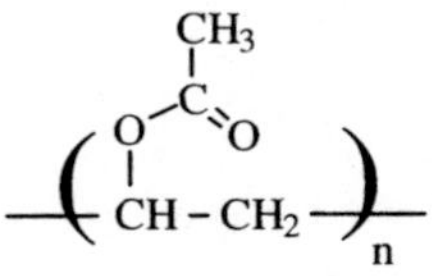 $-(\underset{}{\overset{OC(=O)CH_3}{CH}}-CH_2)_n-$	Poly(vinyl acetate)	latex paints, adhesives
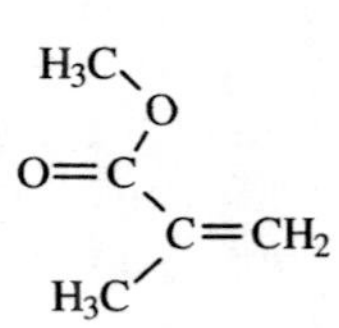 $H_3C-O-C(=O)$\C=CH$_2$ / H_3C	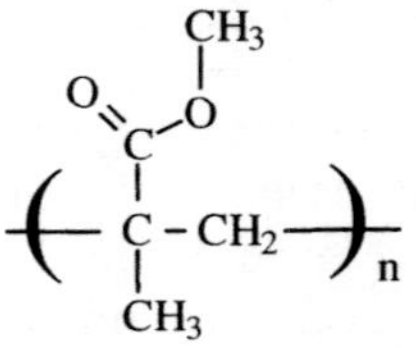 $-(\underset{CH_3}{\overset{C(=O)OCH_3}{C}}-CH_2)_n-$	Poly(methyl methacrylate) (perspex, plexiglass)	window materials, transparent sheeting and glass substitutes
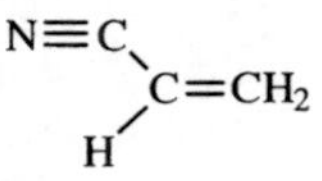 $N\equiv C$\C=CH$_2$ / H	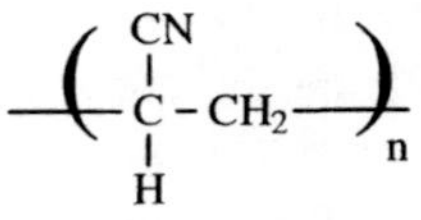 $-(\underset{H}{\overset{CN}{C}}-CH_2)_n-$	Polyacrylonitrile (orlon, acrylan)	fibres, clothing

Organic Hydrocarbons

1.3 AROMATIC HYDROCARBONS

Aromatic hydrocarbons are members of a large and important class of hydrocarbons. The simplest member of the class is benzene with molecular formula C_6H_6 (see Figure 1.1). The molecular formula for benzene suggests a high degree of unsaturation and benzene might be expected to be a highly reactive molecule. In fact, compared to alkenes and alkynes, which are other unsaturated hydrocarbons, benzene is exceptionally stable and unreactive. The stability of benzene and related aromatic hydrocarbons is due to the fact that the π-bonds can be delocalised over a number of atoms, rather than being confined to pairs of carbon atoms as occurs in alkenes.

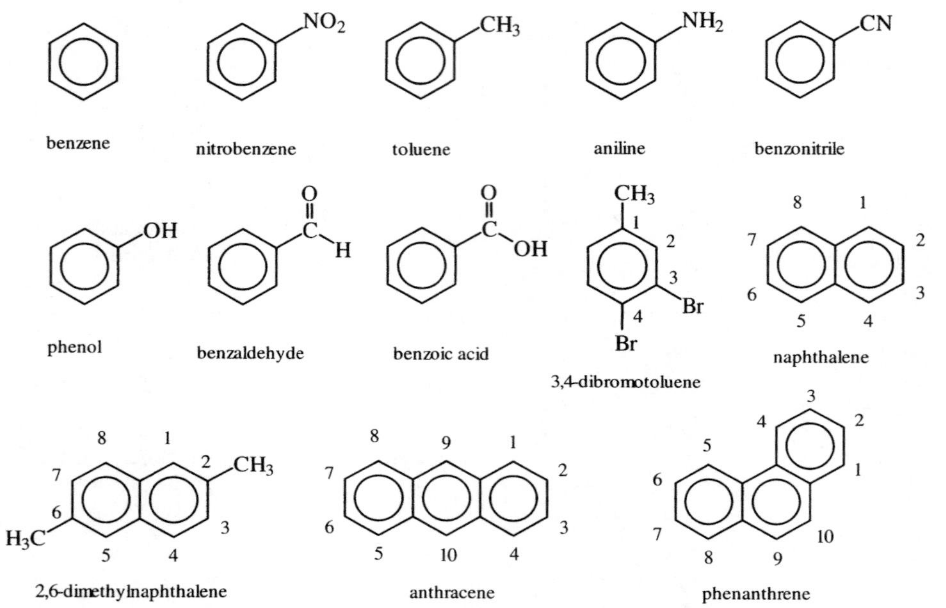

Figure 1.10 Structures, names, and numbering systems of some common aromatic compounds

There is no widely used systematic nomenclature for naming aromatic rings. Each ring system is given a common name and the structural formulas and names of a number of aromatic compounds are shown in Figure 1.10. The aromatic rings are represented by hexagons with a circle inscribed inside to denote aromatic character. In naming derivatives of the aromatic hydrocarbons, it is often necessary to indicate the position on the aromatic ring at which some side chain or other group is attached and the numbers in Figure 1.10 show the common numbering conventions. When there are two groups attached to a benzene ring, there are three possible isomers and these are called the **ortho-**, **meta-**, and **para-**isomers. *Ortho-*, *meta-*, and *para-* can be abbreviated *o-*, *m-* and *p-*.

ortho-dibromobenzene   *meta*-dibromobenzene   *para*-dibromobenzene
o-dibromobenzene   *m*-dibromobenzene   *p*-dibromobenzene
1,2-dibromobenzene   1,3-dibromobenzene   1,4-dibromobenzene

Electrophilic substitution reactions of aromatic hydrocarbons

Although aromatic hydrocarbons are unsaturated, they do not readily undergo addition reactions. The delocalised π-bonding causes aromatic compounds to behave differently to either alkenes or alkynes. For example, benzene does not add Cl_2, Br_2, or H_2 under normal laboratory conditions. Aromatic hydrocarbons do undergo substitution reactions but only with extremely powerful electrophiles. This type of reaction is termed **aromatic electrophilic substitution** and suitably powerful electrophiles include the nitronium ion (NO_2^+), the bromonium ion (Br^+) and the chloronium ion (Cl^+). These electrophiles are too reactive to be stable species and are usually generated *in situ* (within the reaction mixture) immediately before they are required for aromatic substitution.

NO_2^+ is generated in a mixture of concentrated sulfuric and nitric acid; Br^+ is generated by reaction of Br_2 with a Lewis acid (an electron pair acceptor), such as $FeBr_3$; Cl^+ is generated by reaction of Cl_2 with a Lewis acid, such as $AlCl_3$:

$$AlCl_3 + Cl_2 \longrightarrow AlCl_4^- + Cl^+$$
$$\text{chloronium ion}$$

$$FeBr_3 + Br_2 \longrightarrow FeBr_4^- + Br^+$$
$$\text{bromonium ion}$$

$$2H_2SO_4 + HNO_3 \longrightarrow H_3O^+ + NO_2^+ + 2HSO_4^-$$
$$\text{nitronium ion}$$

A CLOSER LOOK: Bonding in Benzene

According to the valence bond view, each of the carbon atoms in benzene is sp^2-hybridised and forms σ-bonds to two neighbouring carbon atoms and a σ-bond to one hydrogen. Each carbon atom has a *p*-orbital which can participate in π-bonding.

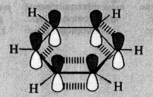

If the bonds were normal C=C bonds, the bonding in benzene could be drawn in either of two identical ways. The *valence bond theory* says that the bonding in benzene is best described as an average or 'hybrid' of the bonding arrangements which can be drawn.

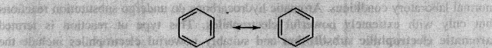

contributors to the hybrid structure of benzene

The bonding in benzene can also be described in terms of *molecular orbital theory*. Here the *p*-orbitals of the six sp^2-hybridised carbon atoms overlap with each other to form a single continuous π-bond. The six electrons in this bond effectively form a toroidal electron cloud which lies above and below the plane of the carbon atoms.

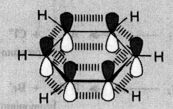

Irrespective of which theory is used to describe the bonding, benzene is a perfect hexagon where all the bond angles are 120° and where all six C—C bonds are identical. The measured bond length in the benzene molecule is 1.39 Å and this is intermediate between the length of a typical C—C single bond (1.52 Å) and a typical C=C double bond (1.33 Å).

When drawing a benzene molecule, it is common to draw a circle in the centre of the hexagon of carbon atoms. This implies 'delocalisation' of the 6 π-electrons around the aromatic ring and indicates the real symmetry of the benzene molecule.

When benzene is treated with bromine in the presence of iron(III) bromide, a hydrogen atom is replaced by a bromine atom:

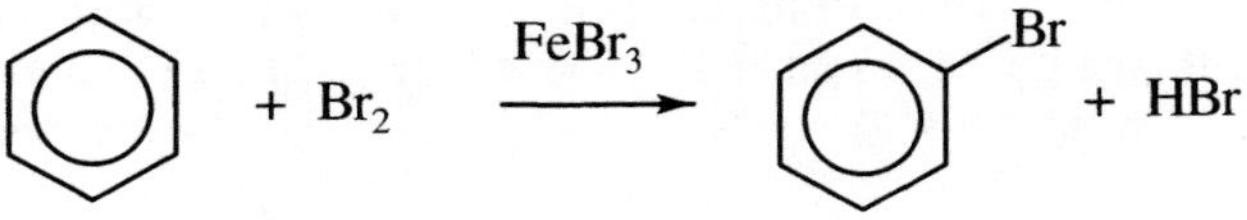

$$C_6H_6 + Br_2 \xrightarrow{FeBr_3} C_6H_5Br + HBr$$

When benzene is warmed in a mixture of concentrated nitric and sulfuric acids, a hydrogen atom is replaced by the nitro group, NO_2:

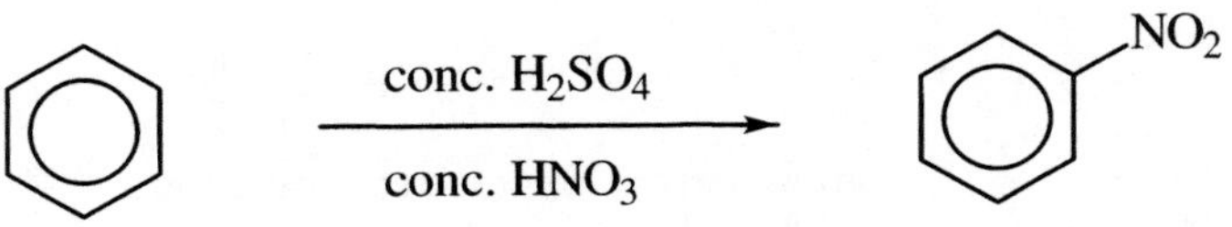

$$C_6H_6 \xrightarrow[\text{conc. HNO}_3]{\text{conc. H}_2\text{SO}_4} C_6H_5NO_2$$

The substitution reactions of aromatic compounds occur via attack of a strong electrophile on the aromatic ring. In the reaction of bromine with benzene in the presence of $FeBr_3$, the mechanism of the reaction can be broken down into a series of steps. The bromonium ion, Br^+, has only six electrons in its valence shell. Using a pair of electrons from the benzene π-system, Br^+ forms a bond to one of the carbon atoms. In doing so it leaves a positive charge on the benzene ring. In an aromatic system the positive charge can be spread onto various carbon atoms in the ring so the cation is best represented with the positive charge not localised on any one carbon atom. Compared to the aromatic compound that we started with, the cation is comparatively unstable and loss of H^+ from the carbon atom to which the bromine is attached converts the molecule back to an aromatic compound. The overall result is that bromine is substituted for hydrogen on the benzene ring:

$$FeBr_3 + Br_2 \rightleftharpoons FeBr_4^- + Br^+$$

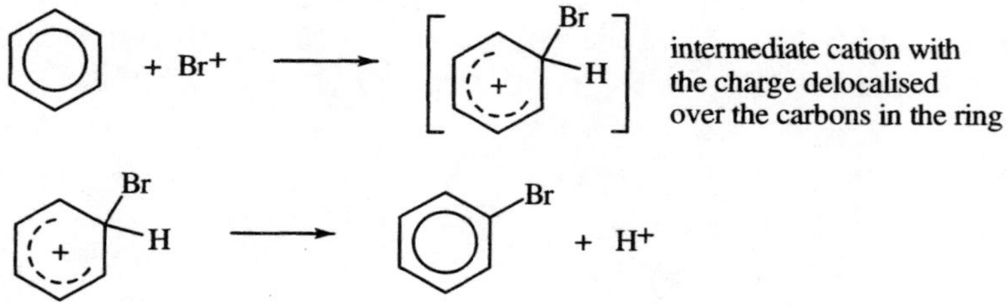

[benzene] + Br$^+$ ⟶ [intermediate cation with Br and H] intermediate cation with the charge delocalised over the carbons in the ring

[intermediate cation] ⟶ [bromobenzene] + H$^+$

Alkyl chains can also be substituted onto aromatic rings in an aromatic electrophilic substitution reaction called a **Friedel—Crafts alkylation** reaction. A carbocation, generated *in situ*, serves as the strong electrophile required to substitute an alkyl group for hydrogen on the ring and it is usually derived from an alkyl halide in the presence of a Lewis acid such as $AlCl_3$:

$$CH_3CH_2Cl + AlCl_3 \longrightarrow CH_3CH_2^+ \; AlCl_4^-$$

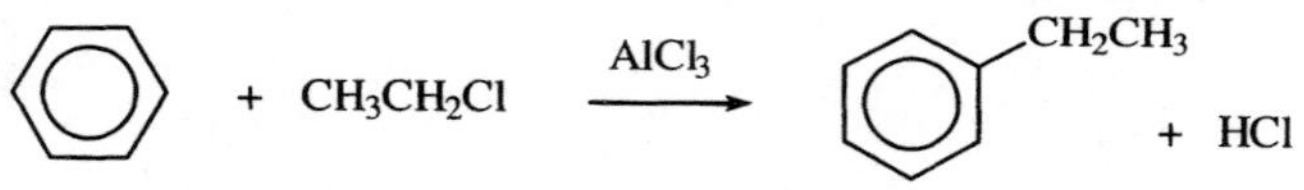

[benzene] + CH_3CH_2Cl $\xrightarrow{AlCl_3}$ [ethylbenzene] + HCl

Aromatic ketones can be formed by a related reaction called a **Friedel—Crafts acylation reaction**. In this case the electrophile is an acylium ion, formed by the reaction of an acid chloride with $AlCl_3$ as a catalyst.

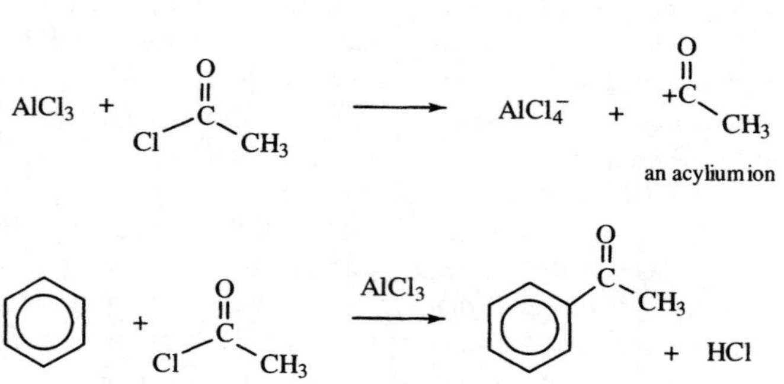

$AlCl_3$ + CH_3COCl ⟶ $AlCl_4^-$ + $^+C(=O)CH_3$ an acylium ion

[benzene] + CH_3COCl $\xrightarrow{AlCl_3}$ [acetophenone] + HCl

A CLOSER LOOK: Aromatic Stabilisation

An estimate of the stabilisation caused by the delocalisation of the electrons in benzene can be obtained by comparing the energy released when hydrogen is added to benzene with the energy released when hydrogen is added to normal alkenes. Hydrogenation of cyclohexene to form cyclohexane is an exothermic process and the reaction releases 120 kJ mol^{-1} of heat:

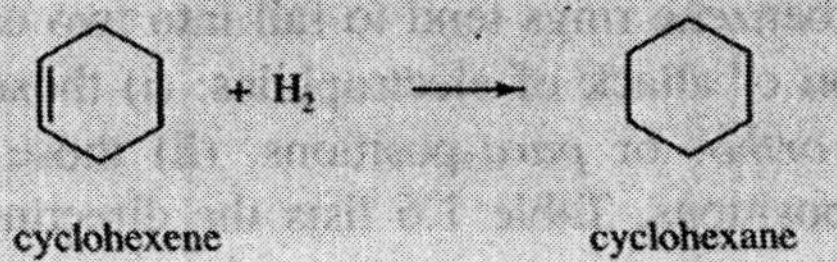

cyclohexene + H$_2$ ⟶ cyclohexane $\Delta H° = -120$ kJ mol^{-1}

Similarly, the heat released on hydrogenating 1,4-cyclohexadiene is 232 kJ mol^{-1}:

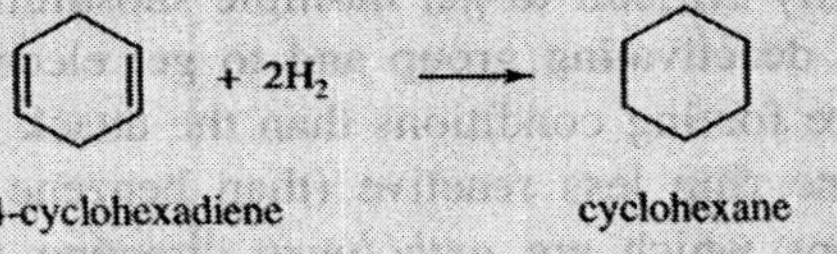

1,4-cyclohexadiene + 2H$_2$ ⟶ cyclohexane $\Delta H° = -232$ kJ mol^{-1}

From these two reactions, it would appear that the heat released on hydrogenating each double bond is about 116 kJ mol^{-1}. There is the equivalent of three double bonds in benzene and it might be expected that the heat of hydrogenating benzene should be about three times -116 kJ mol^{-1}, or -348 kJ mol^{-1}. The hydrogenation of benzene to form cyclohexane is difficult (it requires heat and high pressures). The reaction is exothermic but the experimentally determined enthalpy change in this reaction is only -208 kJ mol^{-1}.

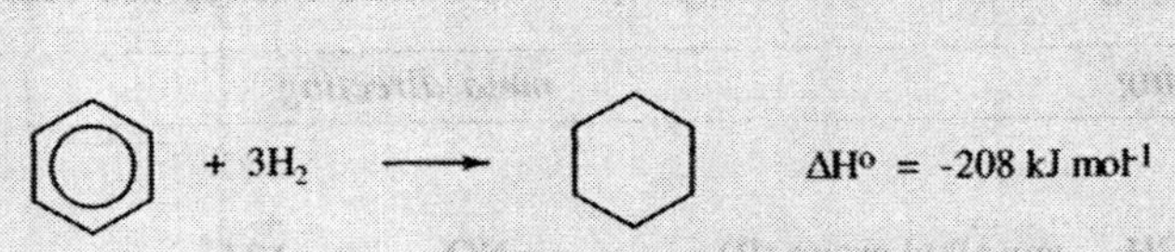

benzene + 3H$_2$ ⟶ cyclohexane $\Delta H° = -208$ kJ mol^{-1}

The fact that the amount of heat released in hydrogenating benzene is less than anticipated, indicates that benzene is more stable than would be expected for a compound which contained three normal double bonds. The difference of 140 kJ mol^{-1} between the expected (-348 kJ mol^{-1}) and the observed (-208 kJ mol^{-1}) heat of hydrogenation is called the *resonance stabilisation energy* and can be attributed to stabilisation of the benzene molecule by the delocalised π-bonding.

Multiple substitutions and the directing influence of groups

It is possible to substitute more than one hydrogen on an aromatic ring with electrophiles. Clearly once the aromatic ring has one group attached, then there are three possible positions (*o*, *m* or *p*) where the next group could be attached. Generally the second (or third or fourth, and so on) substitution does not occur at random, but the groups which are attached to the aromatic ring influence, or direct, the position where the next substitution will occur.

Most groups substituted on benzene rings tend to fall into two categories as far as their ability to direct the position of attack of electrophiles: (i) those which direct the incoming electrophile into the *ortho-* or *para-*positions, (ii) those which direct the incoming attack into the *meta-*positions. Table 1.6 lists the directing effects of some common groups.

In general terms, if there is a C=O group attached directly to the benzene ring, or if the group bears a formal positive charge, then the group will be *meta* directing; otherwise the group is *ortho/para* directing.

Those groups which direct into the *meta* position also tend to make the benzene ring less reactive so it is generally difficult to get multiple substitutions to occur. For example, the —NO_2 group is a deactivating group and to get electrophilic attack on nitrobenzene requires much more forcing conditions than the attack on benzene itself. Groups which make the benzene ring less reactive (than benzene itself) are called *deactivating* groups. Most groups which are *ortho/para* directing tend to make the benzene ring more reactive so it is relatively easy to get multiple substitutions when alkyl groups or —OH groups are present on the benzene ring. Groups which make the benzene ring more reactive (than benzene itself) are called *activating* groups. Halogen substituents are *ortho/para* directing, but deactivate the benzene ring towards further substitution.

Table 1.6 Directing effects of common groups in aromatic electrophilic substitution

ortho/para directing	*meta* directing
—CH_3, —CH_2CH_3, any alkyl group (R)	—NO_2, —$\overset{+}{N}H_3$
—OH, —OCH_3, —OR, —O—C(=O)—R	—C(=O)—R, —C(=O)—H
—NH_2, —NHR, —NR_2	—C(=O)—OH, —C(=O)—OR, —C(=O)—NH_2
—Cl, —Br, —I	

SAMPLE EXERCISE 1.7

Give the product(s) obtained when: (a) benzoic acid is reacted with a mixture of concentrated nitric acid and concentrated sulfuric acid; (b) toluene is reacted with bromine in the presence of $FeBr_3$.

Solution: In (a), a mixture of concentrated nitric and sulfuric acids produces the nitronium ion (NO_2^+) as an electrophile in the reaction mixture. The carboxylic acid group already attached to the aromatic ring will direct the attack of an electrophile to the *meta* position so the product of the reaction will be *m*-nitrobenzoic acid.

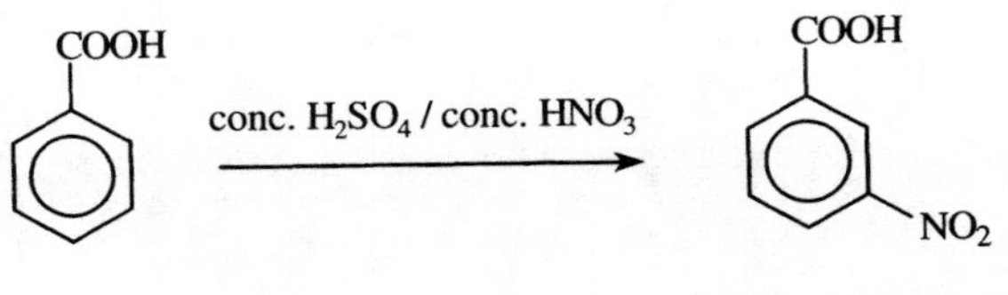

$$\text{C}_6\text{H}_5\text{COOH} \xrightarrow{\text{conc. H}_2\text{SO}_4 / \text{conc. HNO}_3} m\text{-nitrobenzoic acid}$$

In (b), bromine in the presence of $FeBr_3$ produces the brominium ion (Br^+) as an electrophile in the reaction mixture. The CH_3 group attached to the aromatic ring will direct the attack of an electrophile to the *ortho* and *para* positions so the reaction will produce a mixture of *o*-bromotoluene and *p*-bromotoluene.

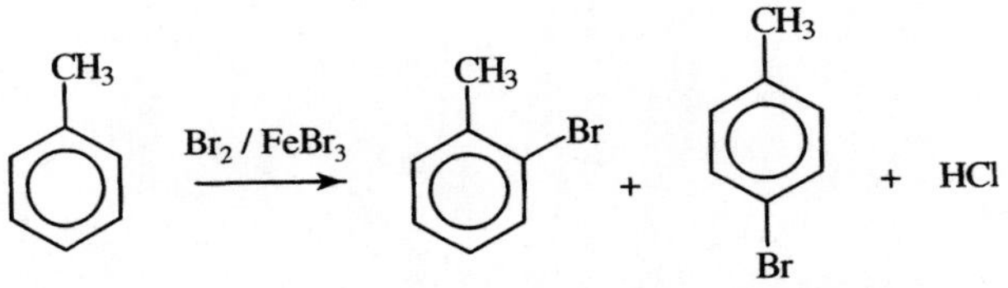

$$\text{C}_6\text{H}_5\text{CH}_3 \xrightarrow{\text{Br}_2 / \text{FeBr}_3} o\text{-bromotoluene} + p\text{-bromotoluene} + \text{HCl}$$

Oxidation of side chains in alkylbenzenes

The presence of an aromatic ring has a significant effect on the reactions of some functional groups. Although C—C bonds are normally resistant to reaction with oxidants, alkyl groups attached to a benzene ring are attacked by oxidising agents such as aqueous $KMnO_4$ or $Cr_2O_7^{2-}$ in the presence of H^+, to give the corresponding carboxylic acids.

$$\text{C}_6\text{H}_5\text{CH}_3 \xrightarrow{\text{KMnO}_4 / \text{H}^+ / \text{H}_2\text{O}} \text{C}_6\text{H}_5\text{COOH}$$

The reaction cleaves any alkyl chain that has at least one hydrogen on the carbon attached to the aromatic ring and converts the alkyl group to a carboxylic acid. Even if the carbon side chain has more than one carbon atom, the group is cut to produce the benzoic acid. The carboxylic acid group is attached to the aromatic ring at the position where the alkyl group was attached. If there is more than one alkyl group attached to the aromatic ring, each is oxidised to a carboxylic acid functional group. This is a useful reaction in determining the position of attachment of alkyl groups on an aromatic ring since the carboxylic acids which result from oxidation can be isolated and compared to authentic samples of aromatic mono-, di-, tri-, and so on carboxylic acids.

Aromatic Compounds Around Us

Benzene was discovered by Michael Faraday in 1825 but it was another 40 years before a realistic structure for the molecule was proposed by Fredrich Kekulé. The compound has been widely used as an industrial solvent, but it is now listed as a known carcinogen and its use has been replaced with *toluene* and *xylene* where possible. *Naphthalene* is the simplest of a series of compounds which have several benzene rings fused together. These molecules are called *polycyclic aromatic compounds*, are isolated from coal tar, and include *anthracene, phenanthrene,* and *pyrene*. Naphthalene is the active component of moth balls.

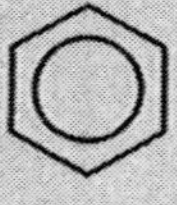

Benzene 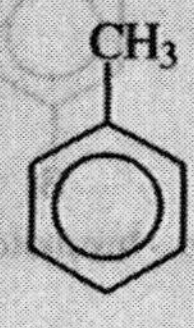   Toluene 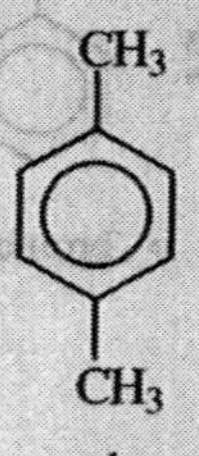   p-xylene 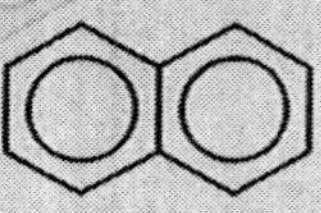   Naphthalene

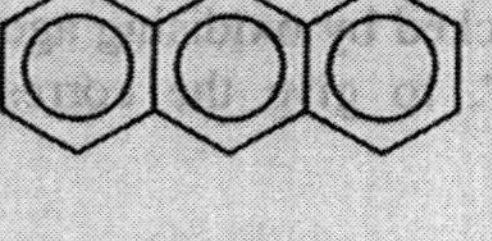

Anthracene 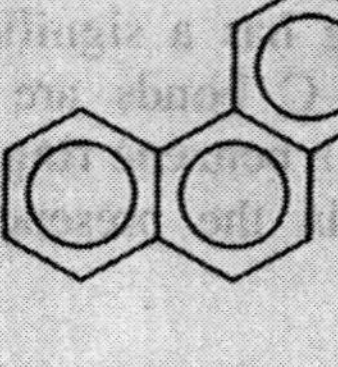   Phenanthrene 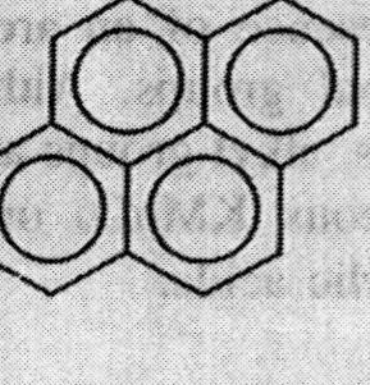   Pyrene

SAMPLE EXERCISE 1.8

Give the condensed structural formulas of the compounds formed when *p*-ethyltoluene is oxidised with an aqueous solution of potassium permanganate.

Solution: $KMnO_4$ reacts with *p*-ethyltoluene to oxidise both of the alkyl side chains to carboxylic acids. The product would be terephthalic acid (1,4-benzenedicarboxylic acid):

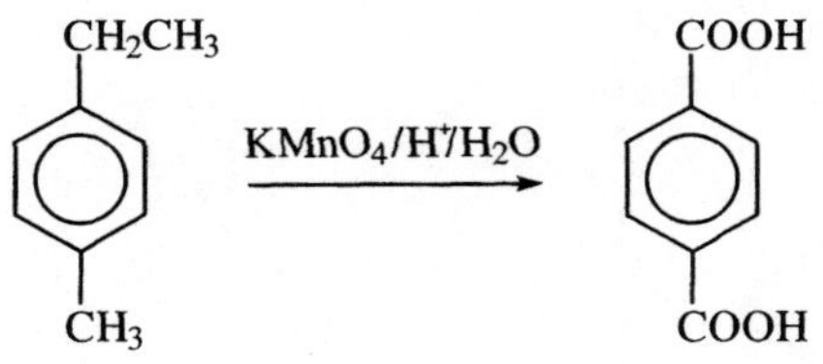

p-ethyltoluene → (KMnO₄/H⁺/H₂O) → terephthalic acid

PRACTICE EXERCISE

Give the condensed structural formulas of the compounds formed when the following compounds are oxidised with an aqueous acidic solution of potassium permanganate:

(a) 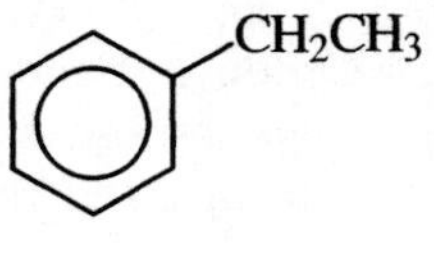 ethylbenzene

(b) 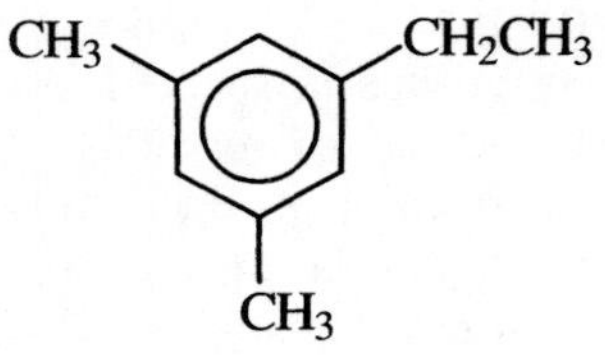 1,3-dimethyl-5-ethylbenzene

(c) 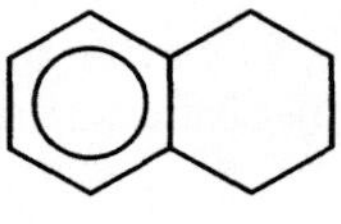 tetralin

Answers:

(a) benzoic acid (benzene with one COOH)

(b) benzene-1,3,5-tricarboxylic acid (benzene with three COOH at 1,3,5)

(c) benzene-1,2-dicarboxylic acid (benzene with two COOH at 1,2)

SUMMARY

Organic compounds are carbon-containing compounds that include both hydrocarbons and derivatives of hydrocarbons. There are four major kinds of hydrocarbons: alkanes, alkenes, alkynes, and aromatic hydrocarbons. *Alkanes* are composed of only C—H and C—C single bonds, and can adopt straight-chain, branched-chain, and cyclic arrangements. *Alkenes* contain one or more C—C double bonds. *Alkynes* contain one or more C—C triple bonds. *Aromatic hydrocarbons* contain cyclic arrangements of sp^2-hybridised carbon atoms bonded with both σ- and π-bonds, and the π-bonds are delocalised.

Isomers are compounds that possess the same molecular formula. *Constitutional isomers* differ in the nature and/or sequence of the atoms or groups. *Stereoisomers* have the same type and sequence of bonds and atoms, but differ in the arrangement of the atoms or groups in space. Stereoisomerism (*cis/trans* or *E/Z* isomerism) is possible in alkenes because of restricted rotation about the C=C double bond. Stereoisomerism (*cis/trans* isomerism) is possible in cyclic alkanes with two or more substituents because substituents can be on the same or opposite faces of the ring and the cyclic structure prevents rotation about the C—C bonds in the ring.

Combustion (burning) of hydrocarbons is an exothermic process and produces CO_2 and H_2O. The chief use of hydrocarbons is as a source of heat energy produced by combustion. Alkanes undergo substitution reactions with halogens (when activated with light) to produce haloalkanes. The *halogenation* of alkanes is difficult to control and mixtures of products are generally obtained.

A *functional group* is a common group of atoms which imparts characteristic chemical and physical properties on an organic compound. Organic compounds are classified and subdivided by the functional groups that they contain.

Hydrogenation of alkenes and alkynes over a platinum or palladium catalyst reduces them to alkanes. The C—C multiple bonds of alkenes and alkynes readily undergo *addition reactions* with chlorine and bromine to give halogenated alkane products. Markovnikov's rule helps to predict the product formed when an unsymmetrical reagent, such as a hydrogen halide (for example, HCl or HBr) is added to a double or triple bond. The *addition of H—X* to an alkene takes place in two steps—the first step involves the reaction of C=C with H^+ to form a *carbocation* and in the second step the cationic centre is attacked by nucleophiles which are present in the reaction mixture. When more than one carbocation can be formed, the major product is derived from the more stable carbocation of the two. *Ozonolysis* is a reaction where alkenes are cleaved by ozone, followed by treatment with a mild reducing agent, to give aldehydes or ketones.

Polymers are large molecules formed by the condensation of small, repeating units called *monomers*. Monomers with double bonds undergo addition polymerisation to form vinyl polymers.

Aromatic hydrocarbons are remarkably stable compounds. Benzene derivatives which have two groups attached to the aromatic ring have three isomers and these are called the *ortho* (*o-*), *meta* (*m-*) and *para* (*p-*) isomers depending on whether the relative positions of the groups are 1,2-, 1,3- or 1,4- respectively.

Aromatic hydrocarbons undergo *aromatic electrophilic substitution* reactions with reagents which are strong electrophiles (bromonium ions, chloronium ions, nitronium ions, carbocations, and acylium ions). The presence of a group on an aromatic ring directs the position of attack of a second electrophile. Groups tend to fall into one of two classes with respect to their *directing influence*: those which direct electrophilic attack into the *meta* position and those which direct electrophilic attack into either the *ortho* or *para* positions. Groups attached to an aromatic ring also influence its ability to undergo electrophilic substitution. Groups which make the aromatic ring easier to substitute than benzene are called *activating substituents* and those that make electrophilic substitution more difficult than in benzene are called *deactivating substituents*. The side chains of alkylbenzenes can be oxidised to convert the alkyl groups to carboxylic acid groups. Side chains are cut at the first carbon so, irrespective of the length or complexity of the alkyl side chain, the product always contains a carboxylic acid functional group attached to the aromatic ring at the position where the alkyl group was attached.

KEY TERMS

addition reaction
addition polymerisation
alkane
alkene
alkyl group
alkyne
aromatic electrophilic substitution
aromatic hydrocarbons
cis/trans
combustion
conformational isomers
constitutional isomers
cycloalkane
directing influence
electrophiles
free radical
Friedel Crafts alkylation

Friedel Crafts acylation
functional groups
gauche, anti, staggered, eclipsed
homologous series
hydrocarbon
Markovnikov's rule
monomer
nucleophiles
1°, 2°, 3° carbocations
ortho, meta, para
ozonolysis
polymers
radical chain reaction
reaction mechanism
stereoisomers
substitution reaction
Z/E

EXERCISES

Hydrocarbon Structures and Nomenclature

1. Give the molecular formula of a hydrocarbon containing five carbon atoms that is:
   (a) an alkane
   (b) a cycloalkane
   (c) an alkene
   (d) an alkyne.
2. Give the molecular formula of an alkane, an alkene, an alkyne, and an aromatic hydrocarbon that in each case contains six carbon atoms.
3. Draw the five constitutional isomers of hexane, C_6H_{14}. Name each compound.
4. Write the condensed structural formulas for all non cyclic alkanes, alkenes, and alkynes that have the molecular formula:
   (a) C_5H_8
   (b) C_5H_{10}
   (c) C_6H_{12}.
5. What are the characteristic bond angles:
   (a) about any carbon atom in an alkane
   (b) about the C—C double bond in an alkene
   (c) about the C—C triple bond in an alkyne?
6. What are the characteristic hybrid orbitals used by:
   (a) a carbon atom in an alkane
   (b) a carbon atom in a double bond in an alkene
   (c) a carbon atom in the benzene ring
   (d) a carbon atom in a triple bond in an alkyne?
7. Write the condensed structural formula for each of the following compounds:
   (a) *(E)*-5-methyl-2-heptene
   (b) 3-chloropropyne
   (c) *ortho*-dichlorobenzene
   (d) 2,2,4,4-tetramethylpentane

(e) 3-ethyl-2-methylhexane (f) 6-chloro-2-methyl-3-heptyne
(g) 1,5-dimethylnaphthalene (h) 1,6-heptadiene

8. Write the condensed structural formula for each of the following compounds:
   (a) 2,2-dimethylpentane (b) 2,3-dimethylhexane
   (c) (Z)-2-hexene (d) methylcyclopentane
   (e) 2-chlorobutane (f) 1,2-dibromobenzene
   (g) methylcyclobutane (h) 4-methyl-2-pentyne.

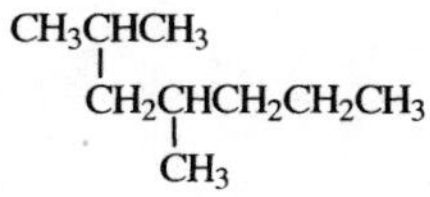

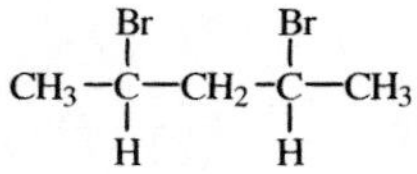

9. Name the following compounds:

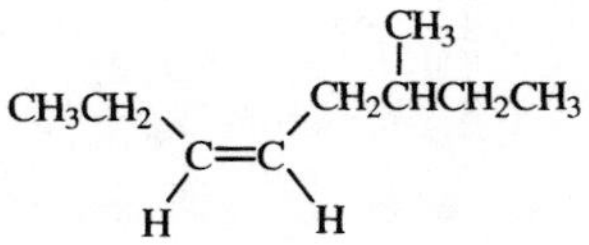

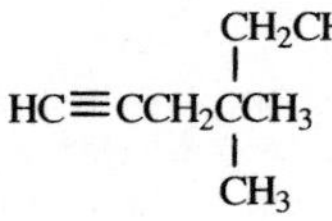

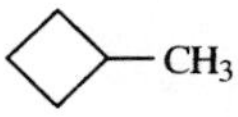

10. Name the following compounds:

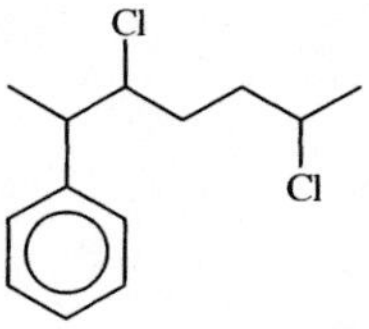

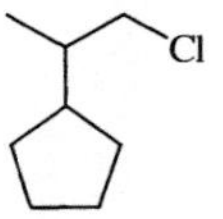

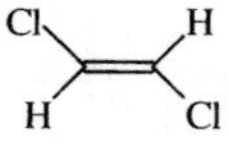

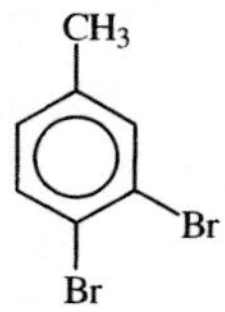

11. Using butene as an example, distinguish between constitutional isomers and stereoisomers.
12. Why is stereoisomerism possible for alkenes but not for non-cyclic alkanes and alkynes?
13. Draw all of the constitutional isomers and stereoisomers of dichloropropene.
14. Indicate which of the following molecules can exist as more than one stereoisomer. For those where more than one stereoisomer is possible, draw their structures:
    (a) 2,2-dichlorobutane
    (b) 2,3-dichloro-2-butene
    (c) 1,3-dimethylbenzene
    (d) 4,4-dimethyl-2-pentyne.

Reactions of Hydrocarbons

15. Give an example, in the form of a balanced equation, of each of the following chemical reactions:
    (a) the combustion of an alkane
    (b) the hydrogenation of an alkene
    (c) an addition reaction of an alkyne
    (d) an electrophilic substitution reaction of benzene.
16. Using condensed structural formulas, write a balanced chemical equation for each of the following reactions:
    (a) hydrogenation of 1-butene
    (b) addition of H_2O to (Z)-2-butene using H_2SO_4 as a catalyst
    (c) combustion of cyclobutane to CO_2 and H_2O
    (d) chlorination of benzene in the presence of $FeCl_3$
17. Using Markovnikov's rule, predict the product formed when excess HCl reacts with each of the following compounds:

    (a) $CH_3CH=CH_2$

    (b) 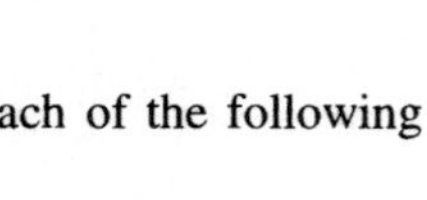 $(CH_3)_2C=C(CH_3)(H)$

    (c) $CH_3C\equiv CH$

18. Using Markovnikov's rule, predict the product of each of the following reactions:
    (a) addition of HBr to $CH_3CH_2CBr=CH_2$
    (b) addition of H_2O, in the presence of H_2SO_4, to

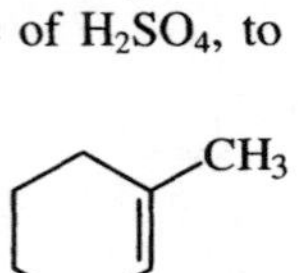

    1-methylcyclohexene (CH_3 substituent on cyclohexene)

    (c) reaction of excess HBr with $(CH_3)_3CC\equiv CH$.

19. When 1-pentene is reacted with HBr, the first species formed is a carbocation. Draw the structures of the two possible carbocations that can be formed in the reaction; indicate which is the more stable carbocation and indicate whether it is a primary, secondary, or tertiary carbocation.
20. When 2-methylpropene is reacted with HBr in the presence of sodium iodide, a mixture of 2-iodopropane and 2-bromopropane is formed. Rationalise this result.
21. Why do addition reactions occur more readily with alkenes and alkynes than with aromatic hydrocarbons?
22. Predict the product (or products) formed in each of the following reactions:

    (a) (Z)-CH$_3$CH=CHCH$_3$ + H$_2$ $\xrightarrow{Pd/C}$

    (b) CH$_3$C≡CH + excess HI $\longrightarrow$

    (c) 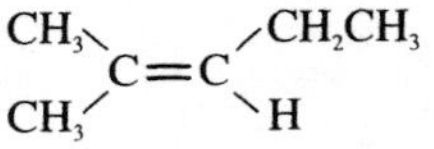 $\begin{array}{c} CH_3 \\ CH_3 \end{array}$C=C$\begin{array}{c} CH_2CH_3 \\ H \end{array}$ + H$_2$O $\xrightarrow{H_2SO_4}$

    (d) CH$_2$=CHCH$_3$ + Br$_2$ $\longrightarrow$

23. Predict the product (or products) formed in each of the following reactions:

    (a) 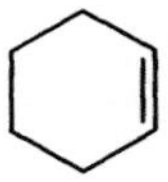 [cyclohexene] + Br$_2$ $\longrightarrow$

    (b) (CH$_3$)$_2$C=CHCH$_3$ + HBr $\longrightarrow$

    (c) CH$_3$CH=CHCH$_2$CH$_3$ + H$_2$ $\xrightarrow{Pd}$

    (d)  [benzene] + CH$_3$CHCH$_3$ (with Cl) $\xrightarrow{AlCl_3}$

24. Predict the product (or products) formed in each of the following reactions:
    (a) benzaldehyde is treated with Br$_2$ in the presence of FeBr$_3$.
    (b) benzoic acid is treated with a mixture of concentrated nitric acid and concentrated sulfuric acid.

25. Write a series of reactions leading to *m*-nitrobenzoic acid, beginning with toluene and using other reagents as needed.
26. The heat of combustion of decahydronaphthalene, $C_{10}H_{18}$, is -6286 kJ mol^{-1}. The heat of combustion of naphthalene, $C_{10}H_8$, is -5157 kJ mol^{-1}. In both cases $CO_2(g)$ and $H_2O(l)$ are the reaction products. Calculate the heat of hydrogenation of naphthalene (ΔH_f° ($H_2O(l)$) = -286 kJ mol^{-1}). Does this value provide any evidence for aromatic character in naphthalene (hydrogenation of one double bond releases approximately 116 kJ mol^{-1})?

Hydrocarbon Derivatives

27. Identify the functional groups in each of the following compounds:

    (a) 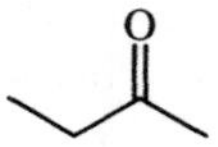 [ketone structure]

    (b) 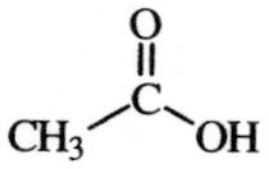 $CH_3-C(=O)-OH$

    (c) [CH₃CH₂CH₂CH₂OH structure]

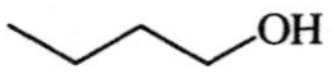

    (d) 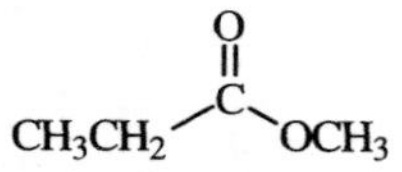 $CH_3CH_2-C(=O)-OCH_3$

    (e) 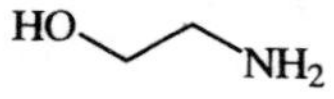 $HO-CH_2CH_2-NH_2$

    (f) $CH_3CH_2NHCH_3$

28. Identify the functional groups in each of the following compounds:

    (a) 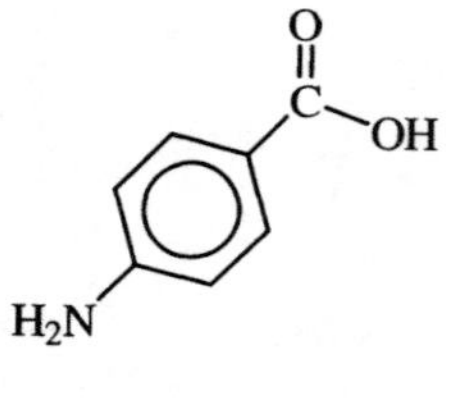 [*p*-aminobenzoic acid structure with H_2N and $-C(=O)OH$ on benzene ring]

    (b) $CH_2=CHCH_2OH$

    (c) 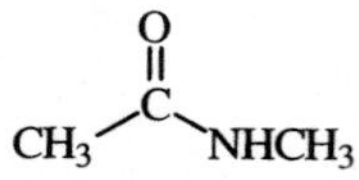 $CH_3-C(=O)-NHCH_3$

    (d)  [acetone structure]

    (e) 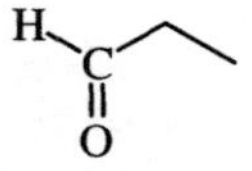 [propanal structure with H-C(=O)]

    (f) $HC\equiv CCH_2OH$

52 The Essentials of Organic Chemistry

Polymers

29. Write the chemical equation for the addition polymerisation of propylene.
30. If two different monomers are involved in addition polymerisation, the resultant polymer is said to be a copolymer of the two monomers. Draw the formula for a copolymer of vinyl chloride (CH_2=CHCl), and 1,1-dichloroethylene (CH_2=CCl_2), the polymer, Saran, used to make food wrapping.

Additional Exercises

31. Draw the structural formulas for two molecules with the molecular formula C_4H_6.
32. Unbranched hydrocarbons are often called straight-chain hydrocarbons. Does this mean that the carbon atoms have a linear arrangement? Explain.
33. What is the molecular formula for:
    (a) an alkane with 20 carbon atoms
    (b) an alkene with 18 carbon atoms
    (c) an alkyne with 12 carbon atoms?
34. Classify each of the following substances as alkane, alkene, or alkyne (assuming that none are cyclic hydrocarbons):
    (a) C_5H_{12}
    (b) C_5H_8
    (c) C_6H_{12}
    (d) C_8H_{18}.
35. Draw the structural formulas for the Z and E isomers of 2-pentene. Does cyclopentene exhibit Z/E isomerism? Explain.
36. Give the IUPAC name for each of the following molecules:

    (a) 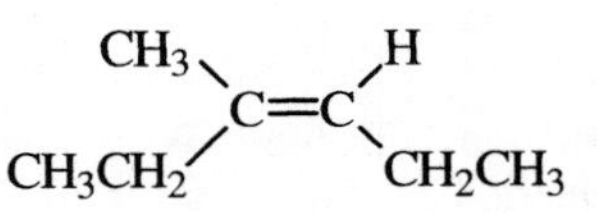 CH_3\   /H
        .... C=C
        CH_3CH_2/   \CH_2CH_3

    (b) 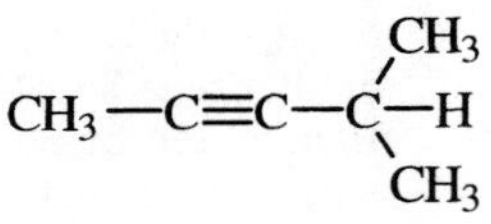 CH_3—C≡C—C(—H)(CH_3)(CH_3)

37. Explain why (E)-1,2-dichloroethylene has no dipole moment, whereas (Z)-1,2-dichloroethylene has a dipole moment.
38. Would you expect cyclohexyne to be a stable compound? Explain.
39. Describe how you could prepare:
    (a) 2,2-dibromopropane from propyne
    (b) 1,2-dichloroethane from ethylene
    (c) 1,2-dibromobutane from 1-butyne.
40. Identify all of the functional groups in each of the following molecules:

(a) 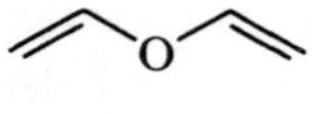 [divinyl ether structure]   an anaesthetic

(b) 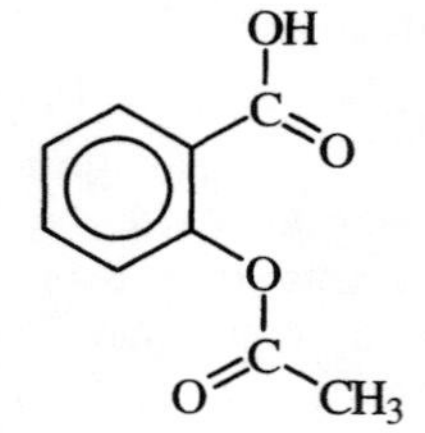 [acetylsalicylic acid structure]   acetylsalicyclic acid (aspirin)

(c) 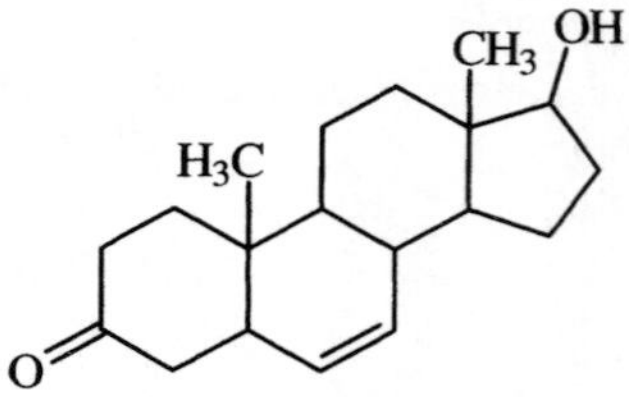 [testosterone structure]   testosterone, a male sex hormone

41. In each of the following pairs, indicate which molecule is the more reactive and give a reason for the greater reactivity:
    (a) butane and cyclobutane
    (b) cyclohexane and cyclohexene
    (c) benzene and l-hexene.
42. Bromination of butane requires irradiation with light. Write a series of reaction steps that can account for the bromination of butane under irradiation conditions.

Derivatives of Hydrocarbons 2

Chapter 1 deals exclusively with hydrocarbons. As important as these compounds are, most organic compounds also contain **heteroatoms** (atoms other than carbon and hydrogen), most notably oxygen, nitrogen, sulfur, phosphorus, and the halogens. In this chapter, the major classes of organic compounds that contain functional groups which include heteroatoms are considered. These functional groups are usually polar and are the most reactive part of the molecule.

2.1   HALOGEN COMPOUNDS

Organic compounds with fluorine, chlorine, bromine or iodine attached to an alkane chain are called **alkyl halides** or *aliphatic halogen compounds*. This term distinguishes alkyl halides from aromatic halogen compounds (also called **aryl halides**) where the halogen atom is attached directly to an aromatic ring. The chemistry of alkyl halides is substantially different from that of aryl halides and, in general, reactions which are well established for alkyl halides *cannot* be directly translated to aromatic halogen compounds.

Naming halogen compounds

Aliphatic halogen compounds are named by firstly identifying the longest carbon chain containing the halogen atom (see Table 2.1). The prefixes *bromo-*, *chloro-*, *fluoro-* or *iodo-* are used to indicate which halogens are present and then a number prefix is used to identify the position of attachment. Aliphatic halogen compounds can be divided into three classes, primary (1°), secondary (2°) and tertiary (3°), depending on how many carbon chains are attached to the carbon bearing the halogen. Alkyl halides are classified as *primary* if there is one carbon chain attached to the C which bears the halogen atom, *secondary* if there are two carbon chains attached to the C which bears the halogen atom, or *tertiary* if the C bearing the halogen has three carbon chains attached. Aromatic halogen compounds are also named by using the prefixes *bromo-*, *chloro-*, *fluoro-* or *iodo-* and the word *benzene* or another aromatic nucleus.

Bonding and properties of alkyl halogen compounds

In terms of their physical properties, alkyl halide compounds are mostly dense liquids and solids that are insoluble in water. The halogens are all more electronegative than carbon and this makes the carbon—halogen bond a polar bond with a slight positive charge ($\delta+$) residing on the carbon end of the bond and a slight negative charge ($\delta-$) on the halogen end:

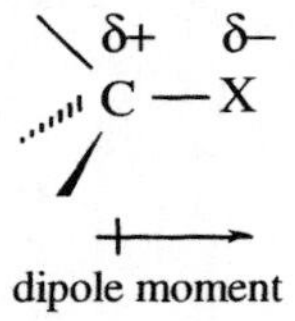

$$\overset{\delta+}{\underset{}{C}}-\overset{\delta-}{X}$$

X = I, Br, Cl, F

dipole moment

Table 2.1 The structures and names of some common organic halides

Structure	Name	Class
CH_3-CH_2-Br	bromoethane	primary alkyl halide
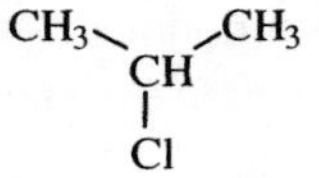 $(CH_3)_2CHCl$	2-chloropropane (or *iso*propyl chloride)	secondary alkyl halide
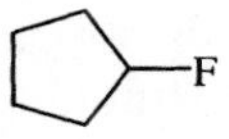 cyclopentyl–F	fluorocyclopentane	secondary alkyl halide
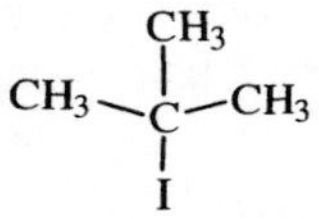 $(CH_3)_3C-I$	2-iodo-2-methylpropane (or *t*-butyl iodide)	tertiary alkyl halide
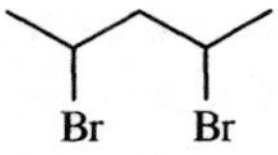 CH₃CHBrCH₂CHBrCH₃	2,4-dibromopentane	two secondary alkyl halides
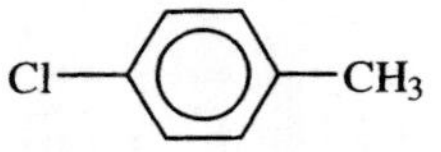 Cl–C₆H₄–CH₃	*p*-chlorotoluene	aryl halide

The strength of the carbon—halogen bond decreases in the order C—F > C—Cl > C—Br > C—I, and the reactions of alkyl fluorides tend to be different from the other alkyl halides, mainly due to the higher strength of the C—F bond. The reactivity of alkyl halides is dominated by the attack of nucleophiles at the carbon atom which bears the halogen atom.

Nucleophilic substitution reactions

Nucleophilic substitution is a reaction which involves the attack of a **nucleophile** (Y^-) at the carbon atom which bears the halogen atom, with displacement of the halide (X^-).

Derivatives of Hydrocarbons **57**

Table 2.2 Nucleophilic substitution

X-R +	Nucleophile (Y⁻)	→	X⁻ +	Product	Product class
" +	HO⁻ (hydroxide)	→	" +	R—OH	alcohol
" +	R'O⁻ (alkoxide)	→	"	R—O—R'	ether
" +	HS⁻ (hydrogensulfide)	→	" +	R—SH	thiol
" +	R'S⁻ (alkanethiolate)	→	"	R—S—R'	sulfide
" +	N≡C⁻ (cyanide)	→	"	R—C≡N	nitrile
" +	R'—C≡C⁻ (alkynide)	→	" +	R—C≡C—R'	longer alkyne
" +	:NH$_3$ (ammonia)	→	" +	R—NH$_3^+$Y⁻	alkylammonium salt
" +	:NR$_3$ (tertiary amine)	→	" +	R—NR$_3^+$Y⁻	tetraalkylammonium salt

This is a general reaction and there are many nucleophiles which can substitute the halogen atom in alkyl halide compounds (Table 2.2). The ease with which the halide can be substituted makes the halogen compounds important intermediates in organic synthesis. A variety of functional groups can be introduced into organic compounds by substitution of the halogen with an appropriate nucleophile in a nucleophilic substitution reaction.

The reactivity of alkyl halides towards nucleophilic substitution depends on the reaction conditions; in general, iodide is more readily substituted than bromide which is more readily substituted than chloride. Alkyl fluorides rarely undergo substitution reactions. The structure of the alkyl halide is also significant—primary alkyl halide compounds tend to have a greater rate of reaction towards nucleophilic substitution than secondary alkyl halides. Tertiary alkyl halide compounds undergo substitution reactions by a different mechanism.

The reaction with tertiary alkyl halides is complicated by competition between substitution and elimination reactions. If the nucleophile used is also a base, for example, hydroxide, then elimination of hydrogen halide from the tertiary alkyl halide to form an alkene can occur. Although the reaction conditions are chosen to maximise the substitution (or elimination) reaction, mixtures of products frequently result.

Not all nucleophiles are anions. The lone pair of electrons on N in molecules such as ammonia and amines, make the nitrogen a good nucleophile and when the amine substitutes the halide of an aliphatic halogen compound, the product bears a positive charge.

Many alkyl halide compounds (particularly methyl halides and reactive primary halogen compounds) are good *alkylating agents*, that is, reagents that will introduce an alkyl group onto a nucleophilic site in a molecule. Such reagents are generally toxic because they alkylate carboxyl groups, amino groups, hydroxy groups, and so on, of biological molecules in living tissue rendering the molecules inactive.

SAMPLE EXERCISE 2.1

Give the condensed structural formula of the compound formed when 1-bromobutane is treated with sodium cyanide.

Solution: Sodium cyanide is a source of the CN⁻ nucleophile and this will undergo a nucleophilic substitution reaction when added to an alkyl halide. The C—Br bond of 1-bromobutane is polarised with a δ+ charge on the carbon and a δ− charge on the bromide. The substitution reaction forms 1-cyanobutane (also called *pentanenitrile*) and liberates bromide ions:

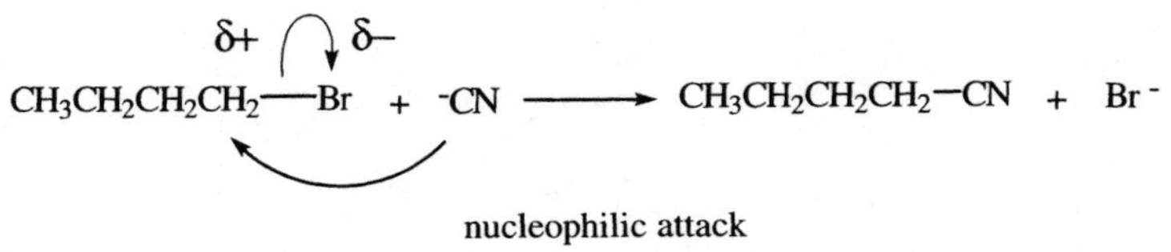

$$CH_3CH_2CH_2CH_2\overset{\delta+}{-}\overset{\delta-}{Br} + {}^-CN \longrightarrow CH_3CH_2CH_2CH_2-CN + Br^-$$

nucleophilic attack

PRACTICE EXERCISE

Give the condensed structural formula of the compounds formed when 1-iodohexane undergoes a nucleophilic substitution reaction using each of the following reagents:
(a) NaOH in dilute aqueous solution,   (b) NaOCH₃ (sodium methoxide),
(c) Na⁺ ⁻C≡C—CH₃ (sodium propynide).

Answers:

(a)  ⌇⌇⌇OH   an alcohol

(b) 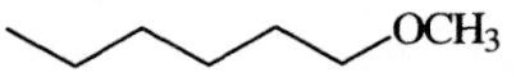 ⌇⌇⌇OCH₃   an ether

(c)   CH₃(CH₂)₄CH₂—C≡C—CH₃   2-nonyne

Aromatic halogen compounds are inert to nucleophilic substitution. If a compound contains a halogen atom bound to an aromatic ring and another halogen atom attached to an alkyl side chain, then the side chain halide can be substituted using the normal conditions for nucleophilic substitution, but the aromatic halogen atom remains intact.

Elimination reactions of alkyl halides

The elimination of HX (X = Cl, Br, I) from an alkyl halide produces an alkene. The **elimination reaction** requires strongly basic conditions and these are usually achieved with bases, such as potassium hydroxide in an alcohol solvent or sodium amide in liquid ammonia solution.

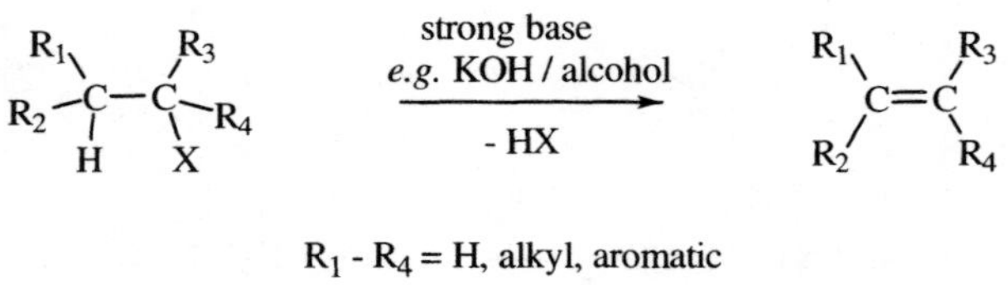

$$\underset{\underset{H}{R_2}}{\overset{R_1}{\diagdown}}C-C\underset{\underset{X}{R_4}}{\overset{R_3}{\diagup}} \xrightarrow[\text{- HX}]{\text{strong base} \atop \textit{e.g.} \text{ KOH / alcohol}} \underset{R_2}{\overset{R_1}{\diagdown}}C=C\underset{R_4}{\overset{R_3}{\diagup}}$$

$R_1 - R_4$ = H, alkyl, aromatic

Note that in some alkyl halides it is possible to form more than one alkene, depending on which carbon the hydrogen atom is removed from. In these cases, a mixture of alkene products is obtained, but it is found that the major alkene formed in an elimination reaction is the one with the more alkyl (or aromatic) groups attached to the sp^2–carbon atoms of the C=C of the product.

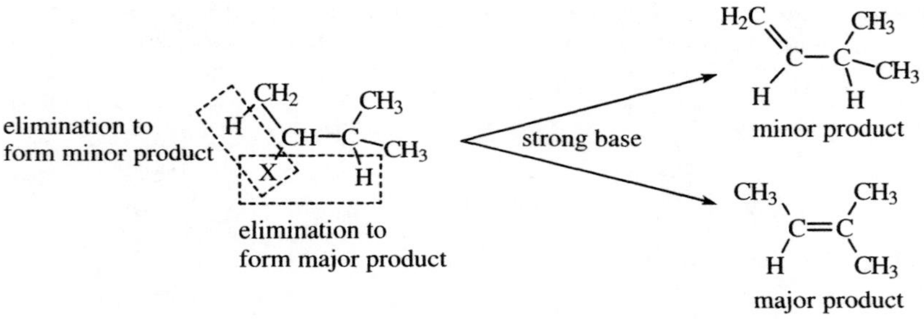

The elimination reaction is a particularly facile reaction with tertiary alkyl halides. The general order of reactivity is that tertiary alkyl halides undergo elimination reactions much more readily than secondary halogen compounds and these undergo reaction more readily than primary halogen compounds. Elimination reactions are the principal cause of difficulty in achieving nucleophilic substitution of tertiary halogen compounds. The course of the reaction, that is, whether products formed arise from a nucleophilic substitution or elimination, can be directed by the careful control of the reagents and the reaction conditions. High concentrations of strong base give predominantly elimination products.

SAMPLE EXERCISE 2.2

2-Bromopropane can react with potassium hydroxide in either an elimination or a substitution reaction. Give (i) the product of the elimination reaction and (ii) the product of the substitution reaction.

Solution: In the elimination reaction OH⁻ acts as a base to remove H⁺ from a carbon adjacent to the one which bears the Br. Elimination of bromide ion would then give propene as a reaction product. With this starting material only one alkene product is possible:

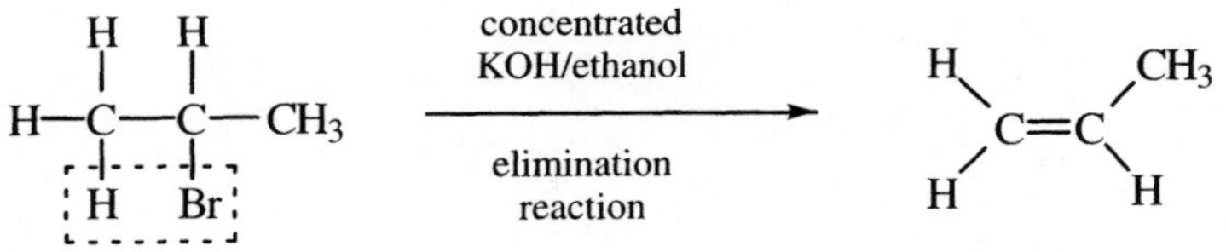

$$\underset{\substack{|||\\ H Br}}{H-\underset{|}{\overset{H}{C}}-\underset{|}{\overset{H}{C}}-CH_3} \xrightarrow[\text{elimination reaction}]{\text{concentrated KOH/ethanol}} \underset{H}{\overset{H}{>}}C=C\underset{H}{\overset{CH_3}{<}}$$

In the substitution reaction, OH⁻ acts as a nucleophile. The carbon atom which bears the Br is polarised δ+ by the presence of the Br and the nucleophile will attack at this carbon to substitute for Br⁻.

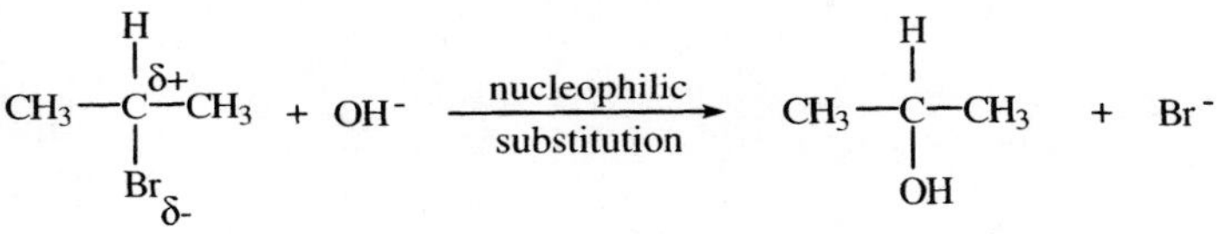

$$CH_3-\underset{\underset{\delta-}{Br}}{\overset{\overset{H}{|}}{\underset{|}{C}}}\!\!^{\delta+}\!\!-CH_3 + OH^- \xrightarrow{\text{nucleophilic substitution}} CH_3-\underset{\underset{|}{OH}}{\overset{\overset{H}{|}}{C}}-CH_3 + Br^-$$

SAMPLE EXERCISE 2.3

2-Iodo-2-methylbutane undergoes an elimination reaction when treated with a hot, concentrated solution of potassium hydroxide in ethanol. Give the structural formulas of the products and indicate which product you would expect to be the major product.

Solution:

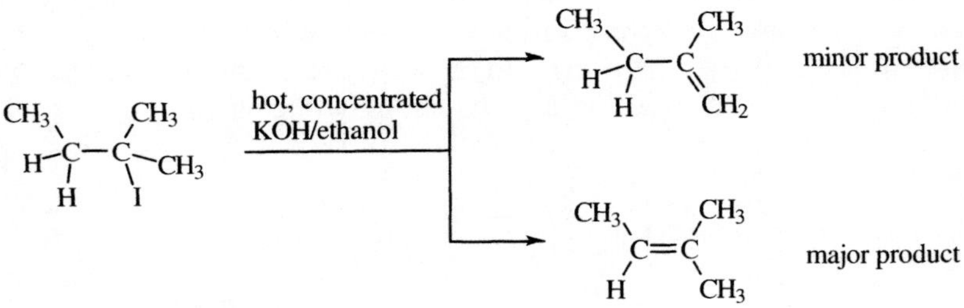

$$\underset{\underset{H}{|}}{\overset{CH_3}{>}}\!C\!-\!\underset{I}{\overset{CH_3}{\underset{|}{C}}}\!-CH_3 \xrightarrow{\text{hot, concentrated KOH/ethanol}} \begin{array}{l} \underset{H}{\overset{CH_3}{>}}\!\underset{H}{\overset{|}{C}}\!-\!\underset{CH_2}{\overset{CH_3}{C}} \quad \text{minor product} \\[2em] \underset{H}{\overset{CH_3}{>}}C=C\underset{CH_3}{\overset{CH_3}{<}} \quad \text{major product} \end{array}$$

Derivatives of Hydrocarbons

PRACTICE EXERCISE

The following alkyl halides undergo elimination reactions when treated with a hot, concentrated ethanol solution of potassium hydroxide. Give the structural formulas of the product(s). If more than one product is possible, indicate which you would expect to be the major product:

(a) 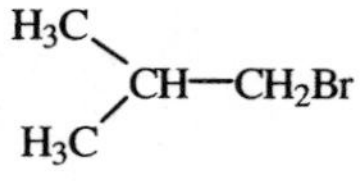 $(CH_3)_2CH-CH_2Br$ (b) 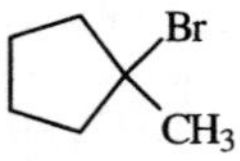 1-bromo-1-methylcyclopentane   (c) 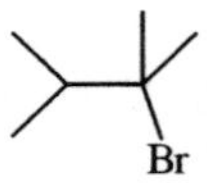 $(CH_3)_3C-CHBr-CH_3$

Answers:

(a) 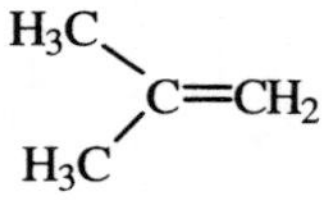 $(CH_3)_2C=CH_2$

(b) 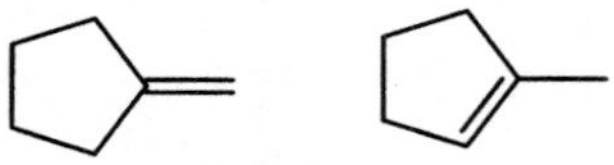 methylenecyclopentane (minor)   1-methylcyclopentene (major)

(c) 2,3-dimethyl-2-butene (major)   3,3-dimethyl-1-butene (minor)

The formation of Grignard reagents

In 1912, the French chemist, Victor Grignard, shared the Nobel Prize in Chemistry with his research supervisor, Philippe Barbier, for the discovery that organic halogen compounds would react with metallic magnesium to give compounds which are extremely important reagents in synthetic organic chemistry. The reagents are called **Grignard reagents** and the formation of the reagent effectively transforms a halogen compound from a species where the carbon which bears the halogen is electrophilic ($\delta+$) to a species where this carbon atom is strongly nucleophilic.

$$R-X \; + \; Mg \xrightarrow{\text{dry ether}} R-Mg-X$$

A Grignard reagent
' $[R]^- [MgX]^{+}$ '

The exact structure of Grignard reagents is not simple and the ether used as a solvent is important in stabilising the reagent as it forms. In practice, a Grignard reagent formed from an alkyl halide R—X, behaves chemically as if it were $[R]^- [MgX]^+$, that is, as a source of R^- ions. Grignard reagents are highly reactive and are normally not isolated as pure compounds. Instead a Grignard reagent is formed in ether solution by reaction of a halogen compound with magnesium metal, and the compound with which you intend it to react is added directly to the reaction mixture.

Both alkyl and aromatic halogen compounds form Grignard reagents. Grignard reagents are named as *organomagnesium salts*; for example, the Grignard reagent formed by reaction of iodomethane with magnesium metal in dry ether solvent is called *methylmagnesium iodide*:

$$CH_3-I \ + \ Mg \ \xrightarrow{\text{dry ether}} \ CH_3-Mg-I$$
$$\text{methylmagnesium iodide}$$

$$C_6H_5-Br \ + \ Mg \ \xrightarrow{\text{dry ether}} \ C_6H_5-Mg-Br$$
$$\text{phenylmagnesium bromide}$$

Grignard reagents are extremely basic compounds and react rapidly with acids, even weak acids. When a Grignard reagent reacts with an acid, the product is the alkane or arene formed by protonating the organic fragment of the reagent. The formation of a Grignard reagent and reaction with an acid is one synthetic method which can be used to remove a halogen from an alkyl or aromatic halide compound. The fact that Grignard reagents react with even weak acids means that the reagents must be formed and manipulated with the scrupulous exclusion of water since any H_2O in the reaction mixture would protonate and decompose the reagent as it formed.

SAMPLE EXERCISE 2.4

Describe a method that you could use to convert iodobenzene to benzene.

Solution: Reaction of iodobenzene with magnesium in dry ether solvent would form the Grignard reagent, phenylmagnesium iodide. Grignard reagents react vigorously with acids and phenylmagnesium iodide would react with any dilute acid to give benzene as the organic product.

$$C_6H_5-I \ + \ Mg \ \xrightarrow{\text{dry ether}} \ C_6H_5-Mg-I \ \xrightarrow{H^+/H_2O} \ C_6H_6$$
$$\text{phenylmagnesium iodide} \qquad \text{benzene}$$

PRACTICE EXERCISE

(a) What are the reagents and conditions required to form cyclopentylmagnesium chloride?
(b) What is the organic product formed when this Grignard reagent reacts with dilute hydrochloric acid?

Answers: (a) Chlorocyclopentane added to magnesium in dry ether solvent, (b) cyclopentane.

Alkyl Halides Around Us

Halothane: 2-bromo-2-chloro-1,1,1-trichloroethane is a volatile, non-flammable compound which has been used as an anaesthetic. Halothane was developed as an anaesthetic to replace diethyl ether which suffered from the drawback that during prolonged exposure, ether could build up to explosive levels in the tissues of some organs of the patient. Modern inhalation anaesthetics are based on halogenated ethers, for example, 'enflurane':

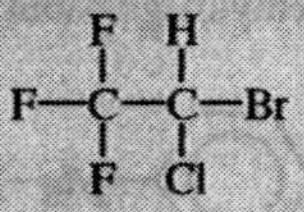

Halothane

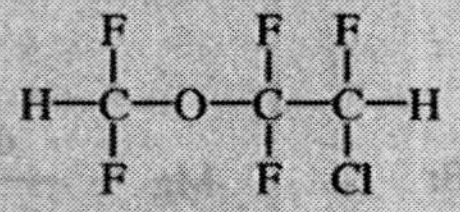

Enflurane

5-Fluorouracil is the active constituent in a chemotheraphy agent used to treat cancer and is the fluorinated analogue of a naturally occurring compound.

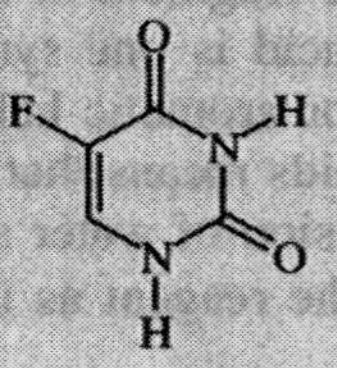

5-Fluorouracil

DDT: Dichlorodiphenyltrichloroethane is one of a class of organochlorine insecticides used very effectively in the 1950s and 1960s to control malaria-carrying mosquitoes, saving an estimated 30 million lives. With time, however, some insects developed a resistance to the chemical and DDT's effectiveness was diminished. Moreover scientists recognised that the inertness of DDT resulted in a persistence of the chemical in the environment. The carbon-chlorine bond is not readily broken down by microorganisms and consequently an accumulation of chlorinated residues in the food chain was observed. The use of DDT is now strictly regulated.

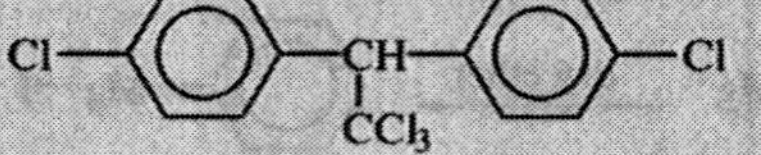

DDT

2.2 ALCOHOLS

Alcohols are organic compounds containing at least one —OH group bonded to an sp^3 hybridised alkyl carbon atom. Compounds which have —OH groups attached to an aromatic ring are called *phenols* and their chemistry is considered in Section 2.3.

Naming alcohols

Alcohols are named by firstly identifying the longest carbon chain containing the —OH group. Derive the name of the alkane and change the *-ane* ending of the alkane name to *-anol* and use a number prefix to identify the position of attachment of the —OH group (Table 2.3). Analogous to the classification of alkyl halides, alcohols are classified as *primary* (1°) if there is one carbon chain attached to the C which bears the —OH functional group, *secondary* (2°) if there are two carbon chains attached to the C which bears the —OH or *tertiary* (3°) if the C bearing the —OH has three carbon chains attached.

Table 2.3 The structures and names of common alkyl alcohols

Structure	Name	Class
CH₃OH	methanol	primary alcohol
CH₃CH₂OH	ethanol	primary alcohol
CH₃CH₂CH₂OH	1-propanol	primary alcohol
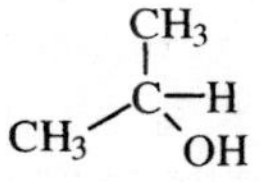 (CH₃)₂CH(OH)	2-propanol or *iso*propanol	secondary alcohol
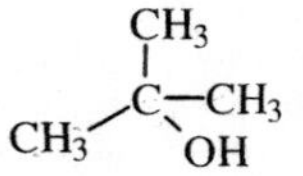 (CH₃)₃C(OH)	2-methyl-2-propanol or *t*-butanol	tertiary alcohol
CH₃CH₂CH₂CH₂OH	1-butanol	primary alcohol
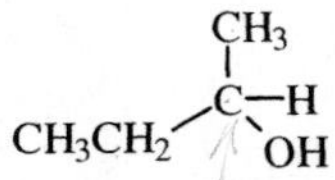 CH₃CH₂CH(CH₃)(OH)	2-butanol or *sec*-butanol	secondary alcohol

Derivatives of Hydrocarbons

PRACTICE EXERCISE

Classify the following alcohols as primary, secondary, or tertiary alcohols:

(a) 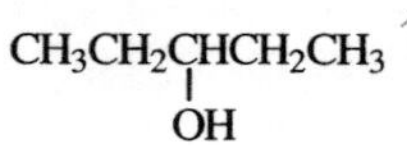 CH$_3$CH$_2$CHCH$_2$CH$_3$
           |
          OH

(b) 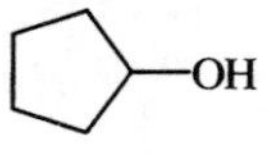 cyclopentyl—OH

(c) 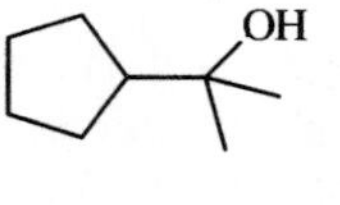 cyclopentyl—C(CH$_3$)$_2$OH

(d) 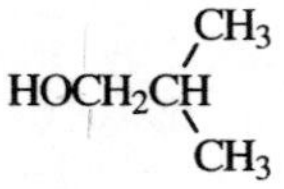 HOCH$_2$CH(CH$_3$)$_2$ (with CH$_3$ groups)

Answers: (a) secondary, (b) secondary, (c) tertiary, (d) primary.

Bonding and properties of alcohols

The oxygen atom of an alcohol is sp^3 hybridised. The oxygen forms a σ-bond to the hydrogen and a σ-bond to an aliphatic carbon. The oxygen atom has two non-bonding pairs of electrons in sp^3 hybridised orbitals:

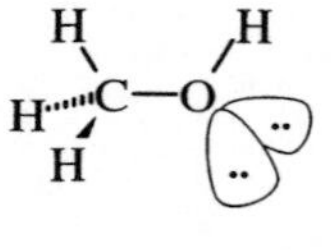

The O—H bond of alcohols is strongly polarised with the H polarised δ+ and the O polarised δ–. The presence of O—H groups allow alcohols to hydrogen bond in much the same way as water molecules. The **hydrogen bond** is a relatively weak interaction between molecules (an *intermolecular* attraction); nevertheless, alcohol molecules are held more tightly together than molecules of non-hydrogen bonding classes of compounds. As a consequence, alcohols have relatively high boiling points compared to other organic compounds of a similar molecular weight.

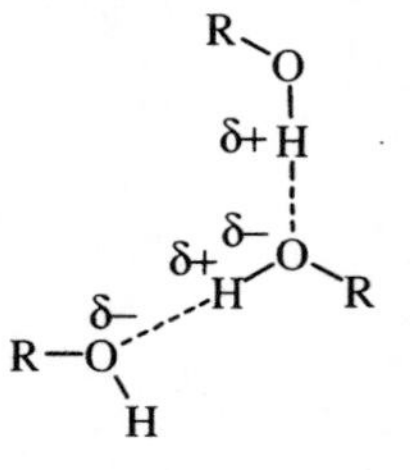

hydrogen bonding between alcohol molecules

As well as hydrogen bonding between alcohol molecules, the —OH groups of alcohols can hydrogen bond with water molecules. This means that alcohols (particularly the lower members of the homologous series of alcohols) are significantly more water soluble than other classes of organic compounds which are not capable of hydrogen bonding. Methanol, ethanol and 1-propanol are all miscible with water; 1-butanol is only partly soluble in water (about 6% v:v at room temperature). As the length of the alkyl chain increases, the alcohols become less water soluble. This reflects a balance between the —OH group, which interacts well with water (the *hydrophilic* part of the alcohol), and the alkyl chain, which interacts poorly with water (the *hydrophobic* part of the molecule).

The formation of alkoxides

Alcohols are very weak organic acids, somewhat weaker than water in their acid strength (about 10^2 to 10^4 times weaker).

$$HO-H \rightleftharpoons HO^- + H^+ \quad pK_a = 15.7$$

$$RO-H \rightleftharpoons RO^- + H^+ \quad pK_a = 16-18$$

If a sufficiently strong base reacts with an alcohol, the hydrogen of the —OH group can be removed as a proton. The hydroxide ion is not sufficiently basic to deprotonate alcohols, but reagents, such as sodium amide ($NaNH_2$), sodium hydride (NaH) or Grignard reagents (RMgX), easily deprotonate alcohols. The RO⁻ ion which results from deprotonation of an alcohol is called an **alkoxide ion**. Alkoxides are named by taking the stem of the name of the alcohol from which they are derived and adding the ending *-oxide*, for example, CH_3O^- derived from methanol is named the *methoxide ion*.

$$CH_3OH + NH_2^- \rightleftharpoons CH_3O^- + NH_3$$

methanol   strong base   methoxide ion

Alkoxides can also be formed by the reaction of sodium or potassium metals directly with alcohols. The reaction liberates hydrogen (usually as bubbles) and is used as a simple diagnostic test for the presence of the —OH functional group.

$$CH_3CH_2CH_2OH + Na \longrightarrow CH_3CH_2CH_2O^- + Na^+ + 1/2\ H_2\ (g)$$

1-propanol   sodium   1-propoxide ion   hydrogen
                 (metal)   (gas)

Alkoxides are strong bases and are frequently used where bases stronger than hydroxide are required or where bases are needed in non-aqueous reactions. Additionally, alkoxides are good nucleophiles (see Section 2.1 and Table 2.2) and react with alkyl halides to give ethers.

SAMPLE EXERCISE 2.5

Give the structure of the organic product obtained when ethanol is reacted firstly with sodium metal and then with 1-bromopropane.

Answer: Alcohols react with sodium metal to form alkoxides (with the liberation of H_2 gas as a byproduct). Alkoxides are good nucleophiles and react with alkyl halides in a nucleophilic substitution reaction. Ethanol reacts with sodium metal to give ethoxide ions which react with 1-bromopropane to give an ether.

$$CH_3CH_2OH \xrightarrow{Na} CH_3CH_2O^- \xrightarrow{BrCH_2CH_2CH_3} CH_3CH_2OCH_2CH_2CH_3$$

ethanol   ethoxide ion   an ether
                             (a nucleophile)

The oxidation of alcohols

Alcohols are oxidised by a variety of oxidising agents. In the laboratory, common oxidising agents include potassium permanganate in either acidic or basic solution ($KMnO_4/H^+$ or $KMnO_4/OH^-$) or potassium dichromate in acidic solution ($K_2Cr_2O_7/H^+$); however there are a great number of reagents that can oxidise alcohols.

The product of alcohol oxidation depends on whether the starting alcohol is a primary, secondary, or tertiary alcohol. Oxidation of primary alcohols proceeds in two steps and the eventual product is a carboxylic acid. Primary alcohols are oxidised initially to aldehydes; however aldehydes are much more easily oxidised than the starting alcohols so they are rapidly oxidised further to give carboxylic acids. Secondary alcohols are oxidised to ketones. Tertiary alcohols are not oxidised by common oxidising agents.

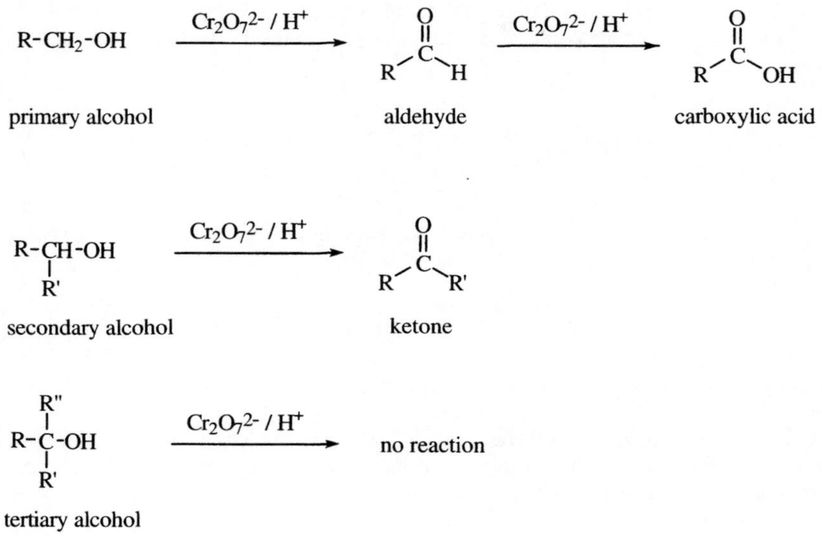

$$R-CH_2-OH \xrightarrow{Cr_2O_7^{2-}/H^+} R-\underset{H}{\underset{|}{\overset{O}{\overset{\|}{C}}}}-H \xrightarrow{Cr_2O_7^{2-}/H^+} R-\underset{OH}{\overset{O}{\overset{\|}{C}}}-OH$$

primary alcohol → aldehyde → carboxylic acid

$$R-\underset{R'}{\underset{|}{CH}}-OH \xrightarrow{Cr_2O_7^{2-}/H^+} R-\overset{O}{\overset{\|}{C}}-R'$$

secondary alcohol → ketone

$$R-\underset{R'}{\overset{R''}{\underset{|}{\overset{|}{C}}}}-OH \xrightarrow{Cr_2O_7^{2-}/H^+} \text{no reaction}$$

tertiary alcohol

Oxidation of methanol is unique among alcohols. Like primary alcohols, methanol is oxidised in stages, firstly to an aldehyde (formaldehyde) which is further oxidised to a carboxylic acid (formic acid). However formic acid can be oxidised further and the eventual products of methanol oxidation are water and carbon dioxide:

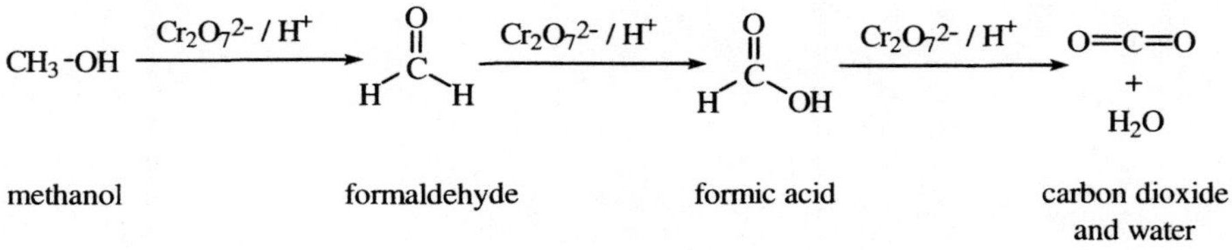

$$CH_3-OH \xrightarrow{Cr_2O_7^{2-}/H^+} H-\overset{O}{\overset{\|}{C}}-H \xrightarrow{Cr_2O_7^{2-}/H^+} H-\overset{O}{\overset{\|}{C}}-OH \xrightarrow{Cr_2O_7^{2-}/H^+} O=C=O + H_2O$$

methanol — formaldehyde — formic acid — carbon dioxide and water

SAMPLE EXERCISE 2.6

Give the structural formulas of the organic products which arise on oxidation of the following alcohols with potassium dichromate in acidic solution: (a) 2-butanol, (b) 3-methyl-1-butanol.

Solution: (a) 2-butanol is a secondary alcohol and is oxidised to a ketone:

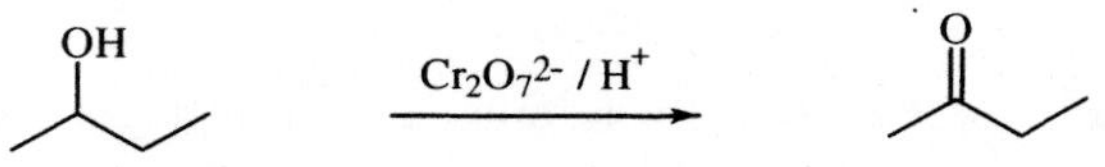

Derivatives of Hydrocarbons **69**

(b) 3-methyl-1-butanol is a primary alcohol and the isolated product of oxidation is a carboxylic acid. An aldehyde is formed as an intermediate compound; however this is oxidised rapidly to the carboxylic acid and cannot normally be isolated as a product in the reaction:

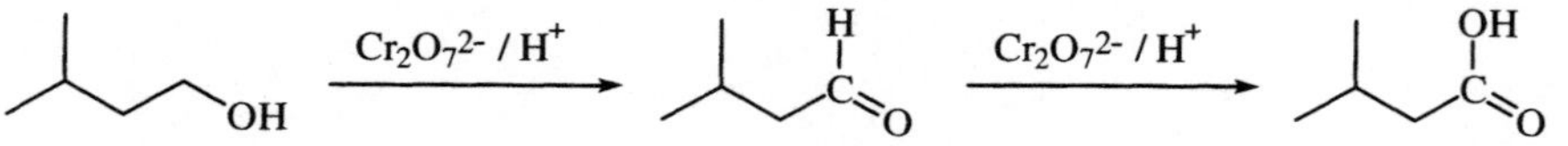

3-methyl-1-butanol (a primary alcohol) $\xrightarrow{Cr_2O_7^{2-} / H^+}$ intermediate aldehyde (not isolated) $\xrightarrow{Cr_2O_7^{2-} / H^+}$ carboxylic acid

PRACTICE EXERCISE

Give the structural formulas of the organic products which arise on oxidation of the following alcohols with potassium dichromate in acidic solution: (a) cyclopentanol (b) 3-methyl-2-butanol (c) 2-methyl-1-butanol.

Answers:

(a) cyclopentanol $\xrightarrow{Cr_2O_7^{2-} / H^+}$ cyclopentanone

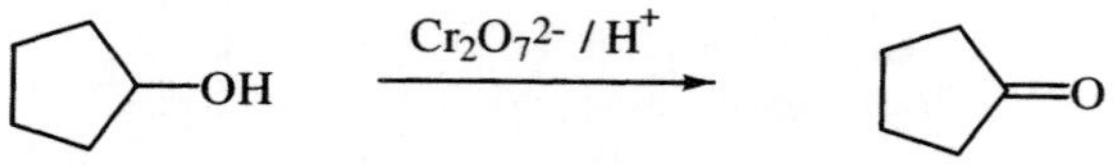

(b) $\xrightarrow{Cr_2O_7^{2-} / H^+}$

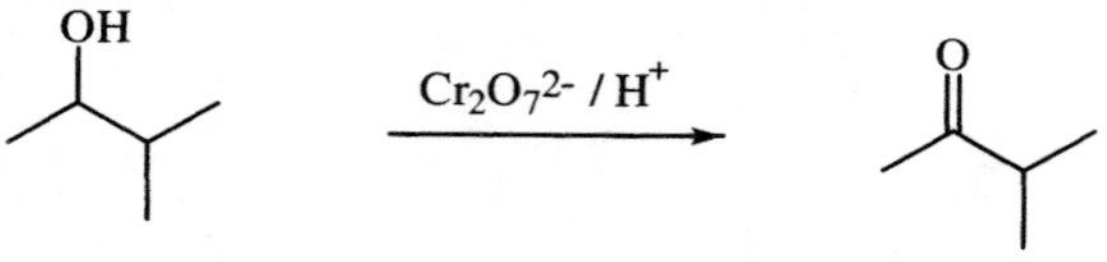

(c) $\xrightarrow{Cr_2O_7^{2-} / H^+}$ COOH

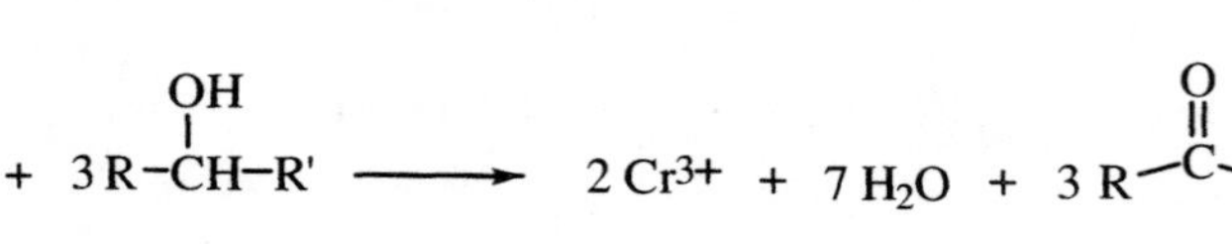

Oxidation of a primary or secondary alcohol with acidified dichromate results in reduction of the dichromate ion to chromium(III):

$$Cr_2O_7^{2-} + 8H^+ + 3\,R\text{-}\underset{\underset{\text{OH}}{|}}{C}H\text{-}R' \longrightarrow 2\,Cr^{3+} + 7\,H_2O + 3\,R\text{-}\overset{\overset{O}{\|}}{C}\text{-}R'$$

Dichromate has an intense orange colour and chromium(III) is blue/green. If an excess of the organic compound is present then a colour change from orange to blue/green is an indication of dichromate reduction and alcohol oxidation. This colour change forms the basis of a useful test for discriminating primary and secondary alcohols, which do undergo oxidation, from tertiary alcohols, which do not.

Conversion of alcohols to alkyl halides

A number of reagents can be used to replace the —OH functional group with —Cl, —Br or —I via a nucleophilic substitution process. The simplest reagents are the concentrated HX acids (X = Cl, Br, or I) which directly convert alcohols to alkyl halides. The reaction takes place in two steps with the first step being protonation of the alcohol with the strong acid. The lone pairs of an sp^3-hybridised oxygen atom are nucleophilic and the oxygen atom is protonated in the presence of strong acids. Protonation effectively converts the —OH group to H_2O which is a much better leaving group and which is much easier to substitute than the —OH group. Substitution of the halide ion for the protonated —OH group affords an alkyl halide. Overall, in this sequence of reactions, the —OH group is activated by H^+ which then allows substitution by the halide nucleophile.

$$R\text{—}OH \xrightarrow{H^+} R\text{—}\overset{+}{O}\begin{matrix}H\\H\end{matrix} \xrightarrow{X^-} R\text{—}X + H_2O$$

$$X = Cl, Br, I$$

$$R\text{—}OH \xrightarrow{HCl} R\text{—}Cl + H_2O$$

$$R\text{—}OH \xrightarrow{HBr} R\text{—}Br + H_2O$$

$$R\text{—}OH \xrightarrow{HI} R\text{—}I + H_2O$$

The reactions work best if the water, which is formed as a reaction product, is removed from the reaction as it is formed. For this reason it is important that the acids HCl, HBr or HI be as concentrated as possible. It is often useful to add a dehydrating reagent to the reaction mixture, for example, molecular sieves or concentrated sulfuric acid which remove water. In this reaction tertiary alcohols tend to react more readily than secondary alcohols and these, in turn, react more easily than primary alcohols.

There are a number of other reagents available to convert alcohols to alkyl halides. The —OH group can be activated and substituted by using a variety of sulfur- or phosphorus-based reagents, including thionyl chloride ($SOCl_2$), phosphorus trichloride (PCl_3), phosphorus pentachloride (PCl_5), and phosphorus tribromide (PBr_3). Thionyl chloride activates the alcohol group by initially forming an oxygen-to-sulfur bond before the carbon-to-oxygen bond is cleaved and the chloride is introduced. Phosphorus tribromide, phosphorus trichloride, and phosphorus oxychloride activate the alcohol group by initially forming an oxygen-to-phosphorus bond before the carbon-to-oxygen bond is cleaved and the halide is introduced.

$$CH_3CH_2CH_2-OH \xrightarrow{SOCl_2} CH_3CH_2CH_2-Cl + SO_2 + HCl$$

$$CH_3CH_2CH_2-OH \xrightarrow{PCl_5} CH_3CH_2CH_2-Cl + POCl_3 + HCl$$

$$CH_3CH_2CH_2-OH \xrightarrow{PBr_3} CH_3CH_2CH_2-Br + HOPBr_2$$

SAMPLE EXERCISE 2.7

Give the condensed structural formula of the product formed when 2-butanol is treated with (a) concentrated hydrobromic acid in the presence of a small amount of concentrated sulfuric acid, (b) thionyl chloride.

Solution: (a) 2-butanol is a secondary alcohol and is converted to 2-bromobutane on treatment with HBr. The presence of H_2SO_4 acts as a dehydrating agent and assists in removing the H_2O byproduct formed in the reaction.

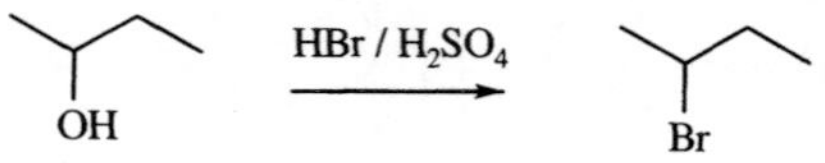

(b) Thionyl chloride will convert the —OH functional group to —Cl so the product would be 2-chlorobutane.

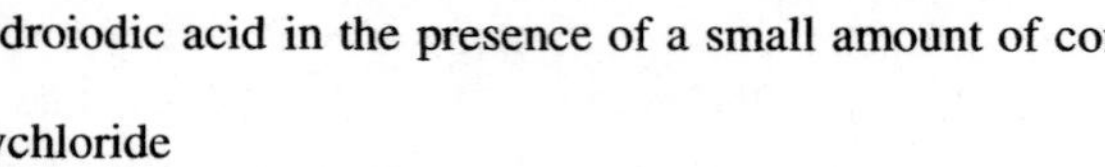

PRACTICE EXERCISE

Give the condensed structural formula of the product formed when cyclohexanol is treated with:
(a) concentrated hydroiodic acid in the presence of a small amount of concentrated sulfuric acid
(b) phosphorus oxychloride
(c) phosphorus tribromide.

Answers: (a) iodocyclohexane, (b) chlorocyclohexane, (c) bromocyclohexane.

Dehydration of alcohols to alkenes

The removal of H_2O from an alcohol is called a *dehydration reaction* and produces an alkene. The removal of water requires a dehydrating reagent and this typically is concentrated sulfuric or phosphoric acid. In this reaction the —OH functional group is removed together with a hydrogen from one of the carbon atoms adjacent to the carbon which bears the —OH group:

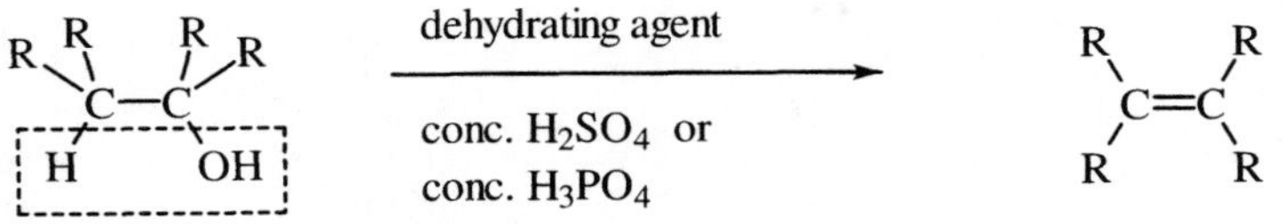

$$\underset{H \quad OH}{\overset{R \quad R}{\underset{|}{C}-\underset{|}{C}}}\overset{R}{\underset{R}{\cdot}} \xrightarrow[\text{conc. } H_3PO_4]{\text{dehydrating agent} \atop \text{conc. } H_2SO_4 \text{ or}} \underset{R \quad R}{\overset{R \quad R}{C=C}}$$

Note that in some alcohols it is possible to form more than one alkene, depending on which carbon the hydrogen atom is removed from. In these cases, a mixture of alkene products is obtained, but the major alkene formed is the one with the more alkyl (or aromatic) groups attached to the sp^2-carbon atoms of the C=C double bond in the product.

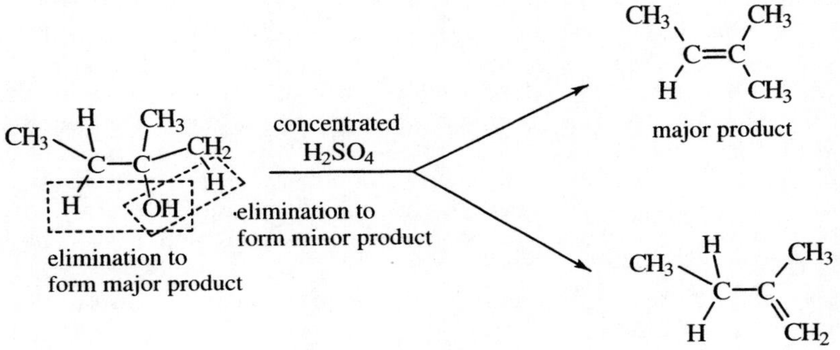

PRACTICE EXERCISE

The following alcohols undergo a dehydration reaction when treated with concentrated sulfuric acid. Give the structural formulas of the product(s). If more than one product is possible, indicate which product you would expect to be the major product.

(a) 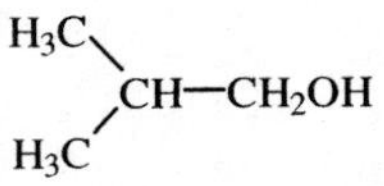 $\underset{H_3C}{\overset{H_3C}{\diagdown}}CH-CH_2OH$ (b) 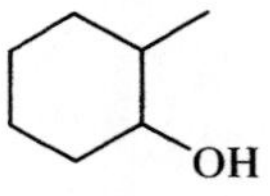 cyclohexane with CH₃ and OH substituents

(c) 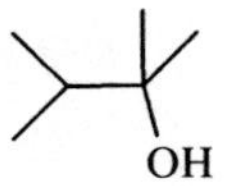 (CH₃)₃C–CH(OH)–CH₃ structure

Derivatives of Hydrocarbons **73**

Answers:

(a) 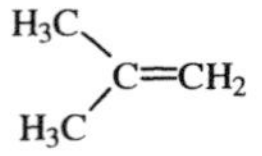 (CH₃)₂C=CH₂ structure: H₃C\C=CH₂ / H₃C

(b) 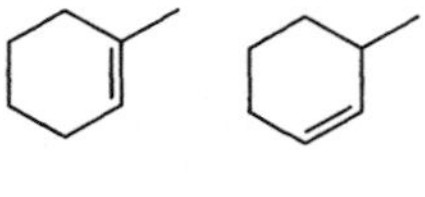 1-methylcyclohexene (major), 3-methylcyclohexene (minor)

(c) 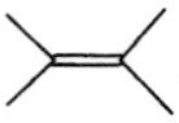 (CH₃)₂C=C(CH₃)₂-type: tetrasubstituted alkene (major), trisubstituted alkene with terminal =CH₂ (minor)

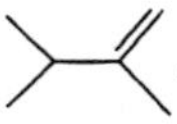

Alcohols Around Us

Ethanol (CH_3CH_2OH): One of the oldest synthetic processes known is the fermentation of sugar and starch to form 'alcohol'. This process is still used for the production of alcohol for drinking where a characteristic taste is derived from the source of the sugars–sugars from grapes are used to produce brandy and wine, and sugars from grain for beer and whisky. Alcohol depresses the central nervous system. The apparent stimulation, which occurs after a small amount of alcohol, results from depression of part of the brain responsible for inhibition and judgment. Increasing amounts of alcohol can affect coordination, balance, and memory. Death occurs with a blood alcohol concentration about 0.4 to 0.6%. Ethanol is rapidly distributed throughout the body, the principal metabolism occurring in the liver where it is oxidised to acetaldehyde and, finally, acetic acid.

The majority of the world's production of ethanol is used as an industrial solvent and is formed by the acid-catalysed hydration of ethylene:

$$CH_2=CH_2 + H_2O \xrightarrow{\text{acid catalyst}} CH_3CH_2OH$$

Vitamin A: An early sign of deficiency of vitamin A is poor night vision and therefore it is no surprise that it is essential for us to see. Vitamin A is a primary alcohol and is formed in the body from the oxidative cleavage of β-carotene, the orange pigment in carrots. While the adage 'carrots help you see in the dark' is true, β-carotene occurs in most green plants and a balanced diet results in adequate supplies of vitamin A to the body.

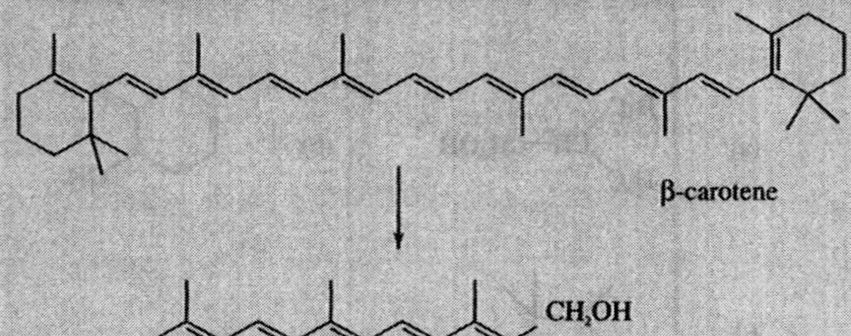

β-carotene

↓

Vitamin A (CH_2OH)

74 The Essentials of Organic Chemistry

> *Terpenes* are natural products which are based on an isoprene unit (2-methyl-1,3-butadiene). Monoterpenes are derived from two isoprene units of which geraniol, found in certain fragrant plants, and menthol, responsible for the flavour and smell of peppermint, are two of many possible examples:
>
> Geraniol   Menthol

2.3 PHENOLS

Phenols are compounds where the —OH functional group is bound directly to an aromatic ring. Although in many ways the reactions of the —OH functional group in phenols parallels that observed in alcohols, there are several important differences which set them apart from alkyl alcohols.

Phenols are much more acidic than alkyl alcohols. Recall that alkyl alcohols have pK_a's in the range 16 to 18 (Section 2.2) which means that they can be deprotonated only with relatively strong bases under non-aqueous conditions. Phenols typically have pK_a's in the range 7 to 11 and the acidity varies widely, depending on the other substituents which are present on the aromatic ring. Acidity in this pK_a range means that, unlike alcohols, phenols react readily with aqueous hydroxide to form **phenoxides**. The increased acidity of phenols compared to that of alcohols can be rationalised by the stabilisation of the negatively charged conjugate base (a phenoxide ion) by delocalisation of the negative charge into the aromatic ring:

phenol + OH⁻ ⇌ phenoxide ion + H₂O

Whereas phenols are generally insoluble (or only slightly soluble) in water, the phenoxide ions formed on deprotonation are water-soluble. Phenols can be separated from a mixture of compounds by extracting the mixture with aqueous base (where the phenolic compounds are soluble) then acidifying the aqueous extracts to precipitate the free phenols.

2.4 ETHERS

Ethers are compounds containing an oxygen atom attached to two alkyl or aryl groups. In ethers, the oxygen is sp^3-hybridised with the oxygen atom, having two lone pairs

Phenols Around Us

Phenol: The simplest of the class of phenols has one hydroxyl group attached to a benzene ring. Phenol used to be called *carbolic acid* reflecting an acidity which is about one million times greater than a saturated hydrocarbon alcohol such as ethanol. When concentrated, phenol is very corrosive towards skin, but it has been used as a bactericide and antiseptic in dilute form since its introduction by Joseph Lister in the nineteenth century. Now hexylresorcinol is more commonly used as a mild antiseptic and disinfectant.

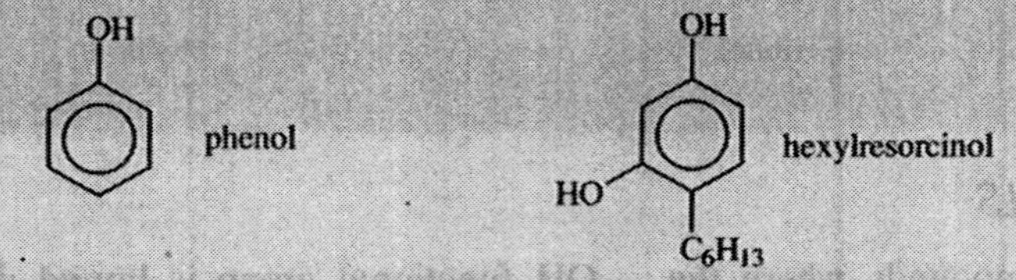

Tyrosine: Tyrosine is a phenol and is required for the correct functioning of our bodies. The compound DOPA is formed from tyrosine and is an intermediate in the formation of the hormone adrenalin (*epinephrine*). Adrenalin is secreted by the adrenal gland when a person is under stress and stimulates the body for a 'flight or fight' response. Parkinson's disease, characterised by an uncontrolled shaking, is associated with a deficiency of dopamine and the symptoms are treated by administering drugs which are converted to dopamine in the body.

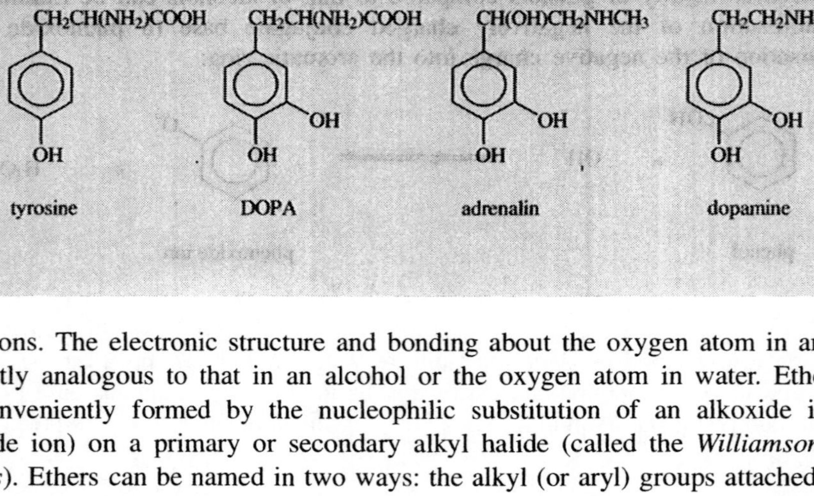

of electrons. The electronic structure and bonding about the oxygen atom in an ether are exactly analogous to that in an alcohol or the oxygen atom in water. Ethers are most conveniently formed by the nucleophilic substitution of an alkoxide ion (or phenoxide ion) on a primary or secondary alkyl halide (called the *Williamson ether synthesis*). Ethers can be named in two ways: the alkyl (or aryl) groups attached to the —O— are named in alphabetical order as two separate words and the word *ether* added. Alternatively, ethers can be named as alkoxy derivatives of alkanes. In this method of naming, the longest continuous alkyl chain forms the stem of the ether name and the alkoxy group is named as a substituent on the alkane backbone.

$$\text{CH}_3\text{CH}_2\text{O}^- + \text{CH}_3\text{—I} \longrightarrow \text{CH}_3\text{CH}_2\text{—O—CH}_3 + \text{I}^-$$

ethoxide ion   iodomethane   ethyl methyl ether
                                                                                      or methoxyethane

If both of the groups attached to the ether oxygen are the same, the ether name is simplified by using the prefix *di-* with the name of the group, for example, CH_3OCH_3 is called *dimethyl ether*.

SAMPLE EXERCISE 2.8

Devise a scheme by which you could synthesise methyl 1-propyl ether using a Williamson ether synthesis.

Solution: The Williamson synthesis reacts an alkoxide with an alkyl halide to make an ether. There are two approaches which can be used to make methyl 1-propyl ether: (i) either methoxide ion is reacted with a 1-halopropane, or (ii) 1-propoxide ion is reacted with a halomethane. In either reaction, the alkoxide ion can be generated by reaction of the parent alcohol with an alkali metal or with a strong base.

$$\text{CH}_3\text{OH} \xrightarrow{\text{Na}} \text{CH}_3\text{O}^- \xrightarrow{\text{CH}_3\text{CH}_2\text{CH}_2\text{Br}} \text{CH}_3\text{CH}_2\text{CH}_2\text{OCH}_3$$

$$\text{CH}_3\text{CH}_2\text{CH}_2\text{OH} \xrightarrow{\text{Na}} \text{CH}_3\text{CH}_2\text{CH}_2\text{O}^- \xrightarrow{\text{CH}_3\text{Br}} \text{CH}_3\text{CH}_2\text{CH}_2\text{OCH}_3$$

PRACTICE EXERCISE

Devise a scheme by which you could synthesise each of the following ethers using a Williamson ether synthesis: (a) 2-butyl methyl ether, (b) anisole (methoxybenzene), (c) cyclopentyl ethyl ether.

Answers: In the first and third case more than one solution is possible, however all require that an alkoxide (or phenoxide) be generated by some means and then reacted with the appropriate halogen compound: (a) react methanol with sodium metal to form sodium methoxide then treat this with 2-iodobutane; (b) react phenol with sodium hydroxide (*note* phenols can be converted to their phenoxides simply by treatment with hydroxide) to form sodium phenoxide then react this with iodomethane (aromatic halides do not readily undergo nucleophilic substitution, so treatment of bromobenzene with methoxide ion will not yield anisole); (c) treat cyclopentanol with a strong base such as sodium hydride to form the cyclopentoxide ion then treat this with bromoethane.

Properties and reactions of ethers

Ethers are remarkable because of their stability and general lack of reactivity. They are excellent solvents for organic solutes and are frequently used as the medium in which other reactions are carried out. Ethers are all water-insoluble and typically very flammable materials. The solvent generally called *ether* is diethyl ether and this is the most commonly used solvent in the organic chemistry laboratory.

Ethers can be cleaved at the carbon–oxygen bonds by concentrated hydroiodic acid (HI). Hydroiodic acid is a very strong acid and reacts by initially protonating the oxygen atom. Attack by I$^-$ (a good nucleophile) breaks one of the carbon—oxygen bonds of the ether to give an alcohol and an organic iodo-compound. The alcohol formed is not isolated since alcohols react with hydroiodic acid to give water and another iodo-compound. Overall, ethers are cleaved by HI to give two organic iodides plus water:

$$R-O-R' \xrightarrow{HI} [R-OH + I-R'] \xrightarrow{HI} R-I + H_2O + I-R'$$

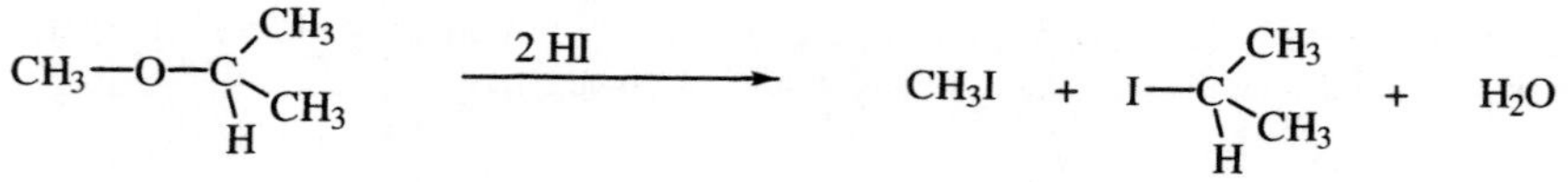

CH$_3$—O—CH(CH$_3$)$_2$ $\xrightarrow{2\ HI}$ CH$_3$I + I—CH(CH$_3$)$_2$ + H$_2$O

methyl isopropyl ether   iodomethane   2-iodopropane

The carbon—oxygen bond between the aromatic ring of an aryl ether is not cleaved by HI. Thus, when one of the groups of an ether is aromatic, the aromatic carbon—oxygen bond remains intact. Diaryl ethers, that is, ethers which contain two aromatic rings attached to the ether oxygen, do not react with HI.

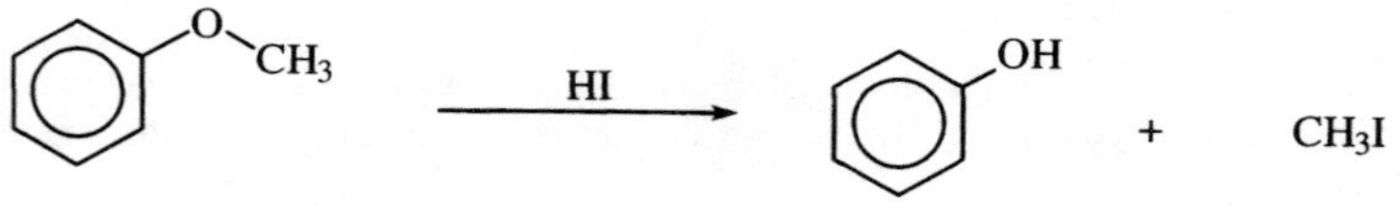

Ph—O—CH$_3$ $\xrightarrow{HI}$ Ph—OH + CH$_3$I

methyl phenyl ether   phenol   iodomethane
(anisole)

Ph—O—Ph $\xrightarrow{HI}$ No reaction

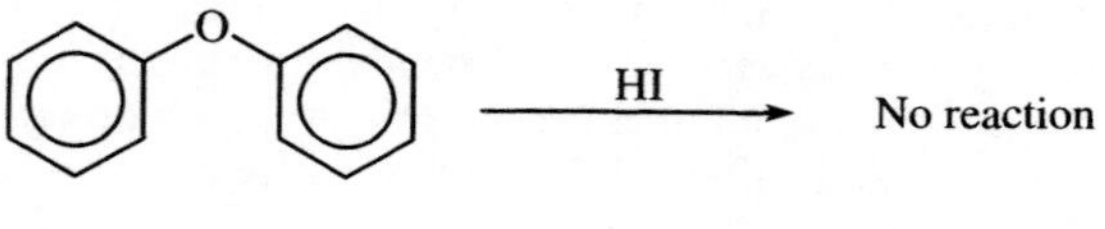

diphenyl ether

Ethers Around Us

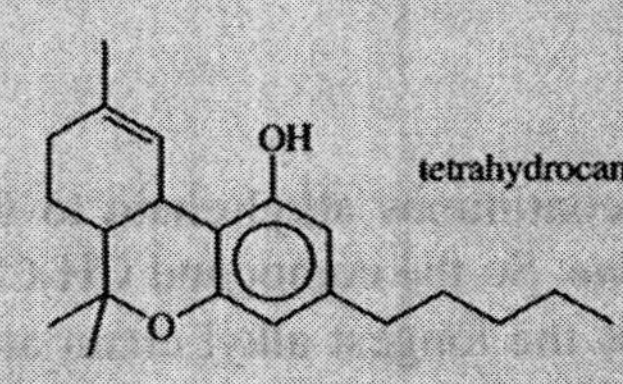

Tetrahydrocannabinol (THC): The hallucinogenic effects of cannabis (marijuana, hashish) are associated with the cyclic ether THC. Another cyclic ether is rotenone which is a natural insecticide and piscicide (fish poison) found in plants of the leguminosae family. It was used in hunting fish by early man.

tetrahydrocannabinol

rotenone

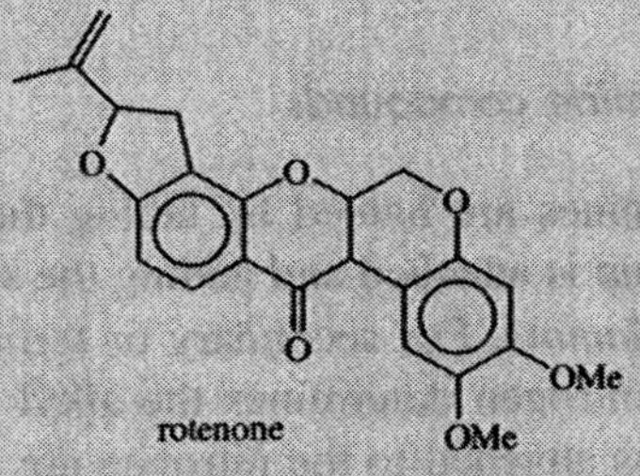

18-Crown-[6]: The importance of crown ethers was recognised by awarding of the 1987 Nobel Prize in Chemistry to Lehn, Cram, and Pederson. These polyethers can wrap around alkali metal ions, increasing the solubility of group 1 salts in organic solvents. The size of the 'crown' can be tailored to the size of the cation radius which makes the complexation selective.

18-Crown-[6]

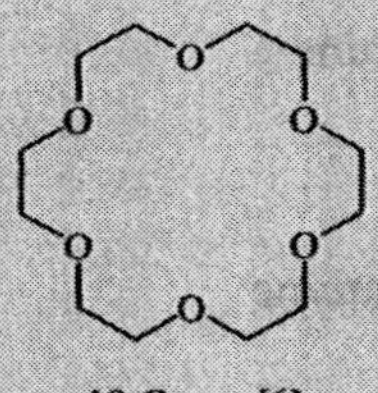

2.5 AMINES

Amines are organic bases and structurally they can be regarded as derivatives of ammonia with one or more of the hydrogen atoms replaced by alkyl or aryl groups. In amines, the nitrogen is sp^3-hybridised with the nitrogen atom having one lone pair of electrons. The electronic structure and bonding about the nitrogen atom in an amine are exactly analogous to that of the nitrogen atom in ammonia:

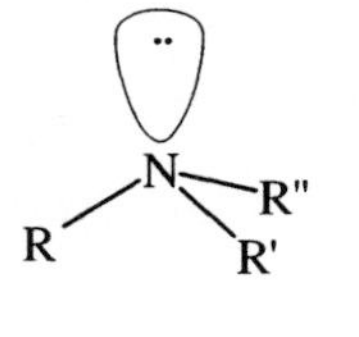

R, R', R" = alkyl, aryl or H

Amines can be divided into three classes: primary, secondary, and tertiary, depending on how many alkyl or aryl groups are attached directly to the nitrogen atom. Amines are classified as *primary* (1°) if there is one carbon chain attached to the amine nitrogen atom; they are called *secondary* (2°) amines if there are two carbon chains attached to the amine nitrogen atom and *tertiary* (3°) amines have three carbon chains attached to the amine nitrogen atom. In addition, tertiary amines can be alkylated by alkyl halides to give *quaternary* ammonium ions.

Naming amine compounds

Primary amines are named by taking the longest continuous alkyl chain to which the —NH_2 group is attached and adding the suffix -*amine*. So the compound $CH_3CH_2NH_2$ is called *ethylamine*. For secondary or tertiary amines the longest alkyl chain attached to the amine nitrogen determines the alkyl group on which the name is based. The other alkyl groups attached to the nitrogen are identified and named in alphabetical order and used as prefixes to the base name, and the italicised letter *N* is used to indicate that the groups are attached directly to the nitrogen. Substituents on carbon atoms are indicated with numbers as described in Section 1.1. Compounds where the amino group is attached directly to an aromatic ring are termed *aryl* or *aromatic amines*.

Structure	Name	Class
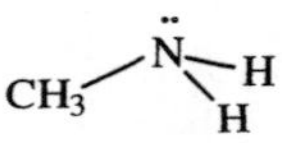 CH_3NH_2	methylamine	primary amine
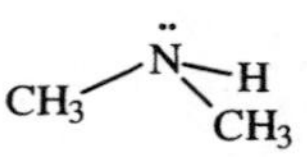 CH_3NHCH_3	dimethylamine	secondary amine
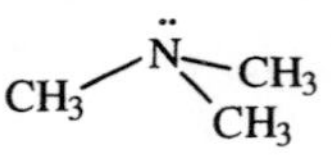 $(CH_3)_3N$	trimethylamine	tertiary amine
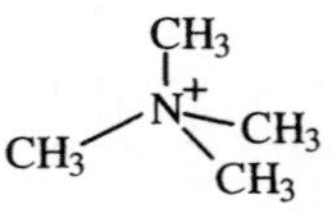 $(CH_3)_4N^+$	tetramethylammonium ion	a quaternary ammonium ion
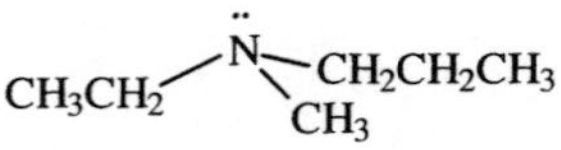 $CH_3CH_2N(CH_3)CH_2CH_2CH_3$	*N*-ethyl-*N*-methylpropylamine	tertiary amine
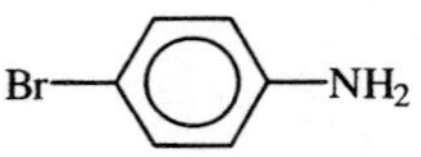 Br-C_6H_4-NH_2	*p*-bromoaniline	primary aromatic amine

SAMPLE EXERCISE 2.9

Name the following amine:

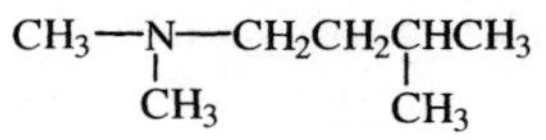

$$CH_3-\underset{\underset{CH_3}{|}}{N}-CH_2CH_2\underset{\underset{CH_3}{|}}{C}HCH_3$$

Solution: There are three non-hydrogen groups attached to the nitrogen so the compound is a tertiary amine. The longest chain has four carbon atoms so the stem name is *butylamine*. The butyl group has a methyl substituent at the third carbon from the nitrogen atom so the name becomes 3-methylbutylamine. Finally there are two methyl substituents attached to the nitrogen atom so the complete name is *N,N*-dimethyl-3-methylbutylamine.

Properties and reactions of amines

Primary and secondary amines have an N—H bond that is strongly polarised with the nitrogen atom being the more electronegative end (δ–) of the bond and the H being the electropositive end (δ+). Like alcohols, there is hydrogen bonding between the molecules of primary and secondary amines, and this leads to amines having generally higher boiling points than comparable compounds without the capacity for hydrogen bonding. The lower members of the amine homologous series are water-soluble because of hydrogen bonding between the water and amine molecules.

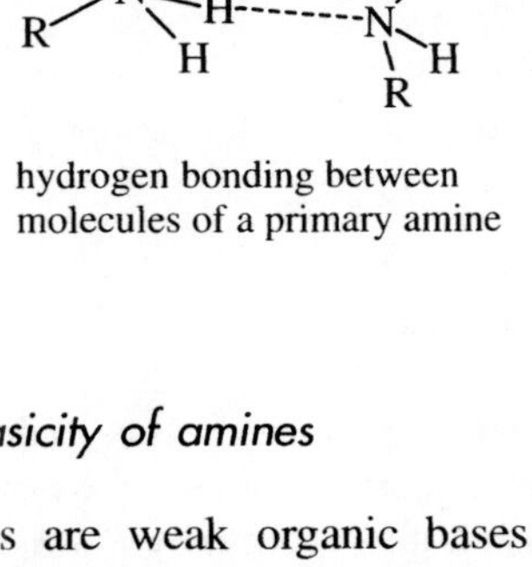

hydrogen bonding between
molecules of a primary amine

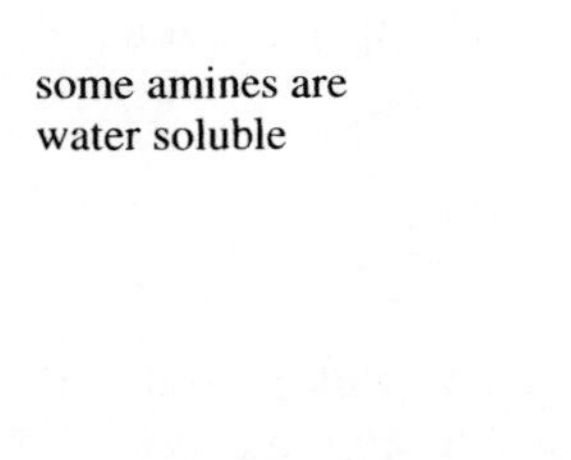

some amines are
water soluble

The basicity of amines

Amines are weak organic bases ($pK_b = 4 - 6$) and of comparable base strength to ammonia. Amines react with water to produce hydroxide ions and hence give a basic solution:

$$R-NH_2 + H_2O \rightleftharpoons R-NH_3^+ + OH^-$$

$$pK_b = [OH^-][RNH_3^+] / [RNH_2]$$

Amines also react reversibly with acids to form substituted **ammonium ions**:

$$CH_3NH_2 \underset{OH^-}{\overset{H^+}{\rightleftharpoons}} CH_3NH_3^+$$

methylamine   methylammonium ion

The protonation of amines to form ammonium ions is easily reversed by the action of bases. Ammonium ions, like other charged species, are generally water-soluble. Apart from the lower members of the series, amines themselves are poorly soluble in water. This means that simply by changing the pH of water from basic to acidic, an amine can be switched from being water-insoluble to water-soluble. In practice, amines can be separated easily from compounds which do not have amine functional groups. Take, for example, a mixture of octane (an alkane) and 1-decylamine (a primary amine). These compounds mix to form a solution and the mixture is insoluble in water. When added to water, a mixture of 1-decylamine and octane forms two phases with the less dense organic phase containing octane and 1-decylamine floating on the more dense aqueous phase. Octane and 1-decylamine can be separated easily from each other by acidifying the water layer. On acidification, the amine reacts with H$^+$ to form the 1-decylammonium ion which is water-soluble and it moves from the organic layer into the water layer. The organic and the aqueous phases can be separated (usually by using a separating funnel which permits the lower layer of a mixture of two immiscible liquids to be drained off). The amine can be recovered from the aqueous phase by adding sufficient base to deprotonate the ammonium ion (Figure 2.1).

Alkylation of amines

Amines are nucleophilies and react with primary and secondary alkyl halides in nucleophilic substitution reactions. This reaction allows primary amines to be converted to secondary amines, secondary amines to tertiary amines, and eventually to quaternary ammonium salts. The reaction of a primary amine with an alkyl halogen compound (for example, iodomethane) will convert it to a dialkylammonium ion. Loss of a proton

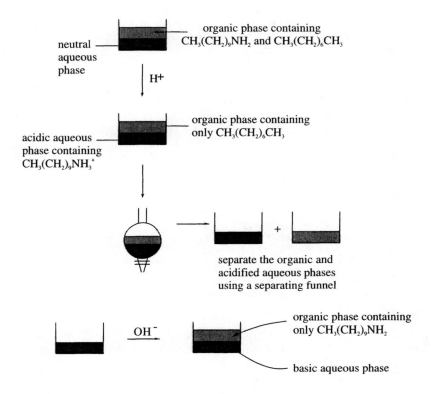

Figure 2.1 A scheme for separating an amine from a hydrocarbon

from the dialkylammonium ion will give a secondary amine. Some of the free secondary amine will always be present from the equilibrium of the ammonium ion with its deprotonated form. Of course any of the secondary amine can react with the alkyl halide present in the reaction mixture to give a trialkylammonium ion. The range of products which can arise from the reaction of an amine with an alkyl halide is difficult to control. Starting from a primary amine, reaction with an alkyl halide could give rise to a mixture of secondary and tertiary amines as well as the corresponding ammonium ions. It is not possible to cleanly alkylate primary (or secondary) amines with alkyl halides to give pure products by this method.

Tertiary amines react with alkyl halides to give quaternary ammonium ions. Once formed, the quaternary ammonium ions are stable. This means that treatment of primary, secondary, or tertiary amines with an excess of an alkyl halide will eventually yield a quaternary ammonium salt following a sequence of alkylation and deprotonation reactions.

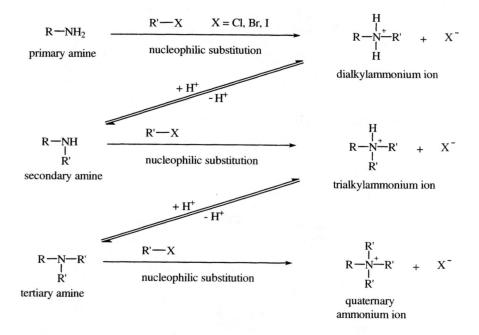

Note that the reaction sequence can be entered at any point (starting with a primary secondary or tertiary amine), but the only place where a clean product can be obtained is at the quaternary ammonium ion stage and this can be ensured if the alkyl halide is present in excess so that the sequence of reactions is naturally driven to completion.

SAMPLE EXERCISE 2.10

Give the condensed structural formula and the name of the product which would arise when N-methylpropylamine is treated with an excess of iodomethane.

Solution: N-methylpropylamine is a secondary amine and would react with iodomethane in a nucleophilic substitution reaction to give (initially) dimethylpropylammonium iodide (which is a salt of the dimethylpropylammonium ion). This is a protonated tertiary amine and is in equilibrium with its deprotonated form, N,N-dimethylpropylamine. The tertiary amine will react with iodomethane (which is present in excess) to produce trimethylpropylammonium iodide which is a quaternary ammonium salt.

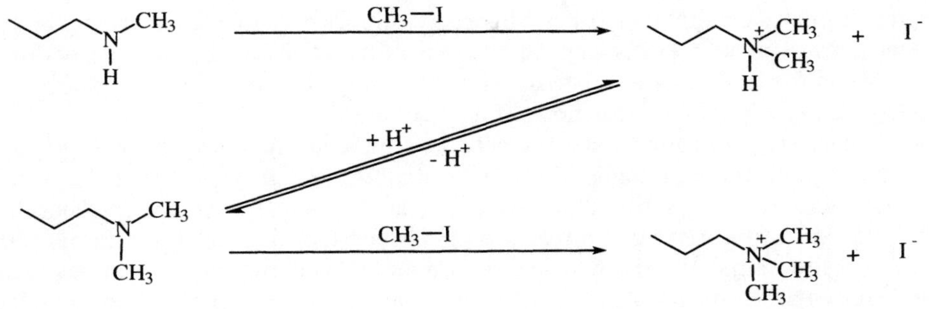

PRACTICE EXERCISE

Give the condensed structural formula and the name of the product which would arise when each of the following amines is treated with an excess of iodoethane: (a) butylamine, (b) aniline, (c) N-methylbutylamine.

Answers:
(a) butyltriethylammonium iodide

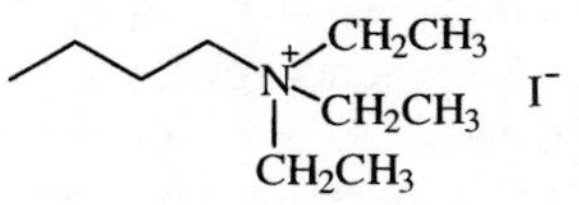

(b) triethylanilinium iodide

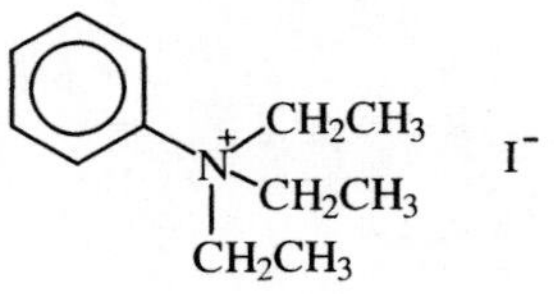

(c) butyldiethylmethylammonium iodide

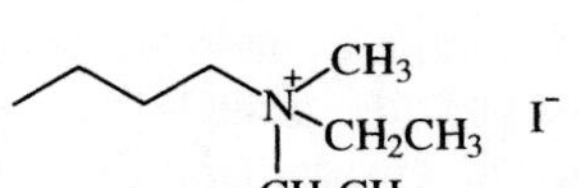

Aromatic amines

Aromatic amines (derivatives of aniline) are bases similar to alkyl amines. In general, aromatic amines are slightly less basic than alkyl amines because the lone pair of electrons on the nitrogen atom interacts with the π-system of the aromatic ring (with delocalisation) and this makes it less readily available for bonding. However, aromatic amines still undergo all of the common reactions of alkyl amines.

Synthesising aromatic amines is easily done by reducing aromatic nitro compounds. The nitro group can be introduced easily as a substituent on an aromatic ring using an aromatic electrophilic substitution reaction (see Section 1.3). Once introduced, the aromatic nitro functional group reacts with various reducing agents to form aromatic amines. Most commonly the nitro group is reduced with either hydrogen gas (using a platinum catalyst) or alternatively with tin, zinc, or iron metal in an acidic solution. With the metal/acid reductions, the nitro group is reduced and the amine which results is protonated in the acid reaction mixture, and the medium must be made basic to release the free amine.

The formation of diazonium ions and their reactions

Aromatic diazonium salts $[Ar-N_2]^+ X^-$ are formed by the reaction of *primary aromatic amines* with *nitrous acid* (HNO_2). These salts are extremely useful synthetic intermediates since there are a number of reagents (generally nucleophiles) which displace the N_2 group (liberating free nitrogen gas) to form a substituted aromatic ring. Nitrous acid is an extremely unstable acid and is usually made *in situ* (in the reaction mixture just before it is required) by reaction of sodium nitrite ($NaNO_2$) with aqueous acid at low temperature (0°C to 5°C).

Reaction of an aromatic diazonium salt with potassium iodide or sodium iodide substitutes iodide for N_2 and this provides a method for introducing an iodine substituent onto an aromatic ring. Reaction of an aromatic diazonium salt with cuprous chloride (CuCl) or cuprous bromide (CuBr) substitutes the N_2 group with chloride or bromide respectively. The reaction of aromatic diazonium salts with copper(I) halides is called the *Sandmeyer reaction* after Traugott Sandmeyer (a Swiss chemist) who discovered

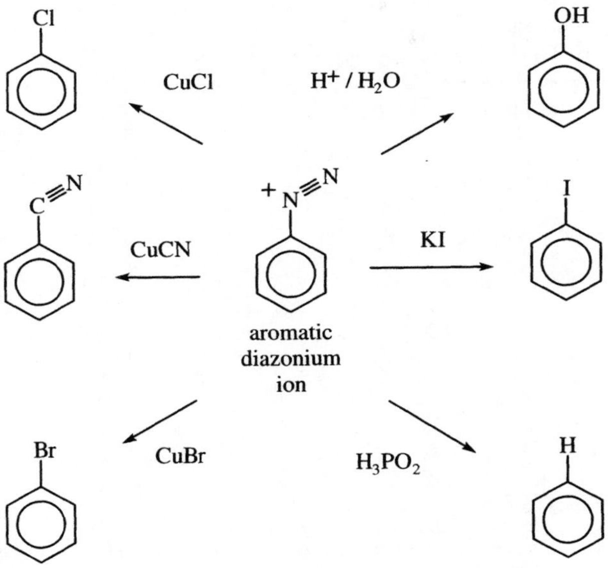

Figure 2.2 Substitution of nitrogen of diazonium ions

the reaction in the late nineteenth century. Reaction of an aromatic diazonium salt with cuprous cyanide (CuCN) substitutes the N_2 group with cyanide and provides a method for forming aromatic nitriles. The N_2 group can be removed and replaced by a hydride by reduction of the aromatic diazonium salt with hypophosphorous acid (H_3PO_2) so this is one method for removing an $-NH_2$ (or $-NO_2$) group from an aromatic ring. The N_2 group can be replaced by —OH to make phenols by reacting the aromatic diazonium salt with hot, aqueous acid.

The ease with which an amine group can be introduced into an aromatic ring (via the nitro group) and the range of reactions the aromatic diazonium ion can undergo, make diazonium salts very useful synthetic intermediates (Figure 2.2).

SAMPLE EXERCISE 2.11

Devise a synthetic scheme for converting benzene to benzonitrile.

Solution: Benzonitrile has a $-C\equiv N$ group attached to a benzene ring. Working backwards, the last step in the sequence should be the reaction of a diazonium salt with CuCN since this is the most convenient way to introduce a nitrile group onto the benzene ring. The precursor to the diazonium salt is the aromatic amine (aniline) and this in turn is best derived from nitrobenzene. So the overall reaction sequence is: (i) nitration of benzene to form nitrobenzene, (ii) reduction of the nitro group to the amino group, (iii) formation of the diazonium salt, (iv) reaction of the aromatic diazonium salt with CuCN. Note that this is a general reaction sequence which can be used to introduce many functional groups onto the aromatic ring.

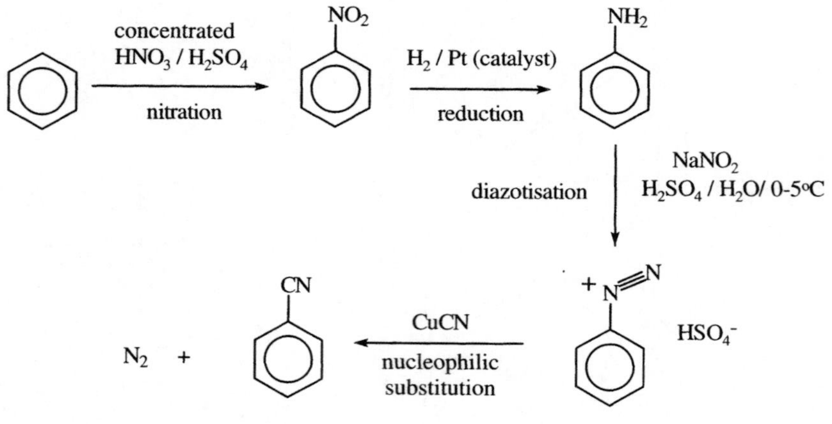

[Reaction scheme: benzene → (concentrated HNO₃/H₂SO₄, nitration) → nitrobenzene → (H₂/Pt catalyst, reduction) → aniline → (NaNO₂, H₂SO₄/H₂O/ 0–5 °C, diazotisation) → benzenediazonium HSO₄⁻ → (CuCN, nucleophilic substitution) → benzonitrile + N₂]

PRACTICE EXERCISE

Devise synthetic schemes for performing the following conversions using diazonium salts as intermediate compounds in the syntheses: (a) *p*-bromoaniline to *p*-bromochlorobenzene, (b) 2-nitrotoluene to toluene, (c) 4-nitrotoluene to *p*-cresol (*p*-hydroxytoluene).

Answers: (a) Convert *p*-bromoaniline to its diazonium salt then treat with CuCl. (b) reduce the nitro group to the amino group then convert the 2-aminotoluene (*o*-toluidine) to its diazonium salt and treat with H₃PO₂. (c) reduce the nitro group to the amino group then convert the 4-aminotoluene (*p*-toluidine) to its diazonium salt and treat with aqueous acid.

Aromatic diazonium salts also undergo coupling reactions with suitably reactive aromatic compounds. Diazonium coupling reactions are aromatic electrophilic substitution reactions in which the aromatic diazonium ion acts as the electrophile. The substrates are usually aromatic compounds which are highly activated and very susceptible to electrophilic attack (usually phenols or aromatic amines). The electrophilic attack of the diazonium ion on the phenol or aromatic amine occurs predominantly in the positions *para* and *ortho* to the activating hydroxy or amino functional groups (see Section 1.3).

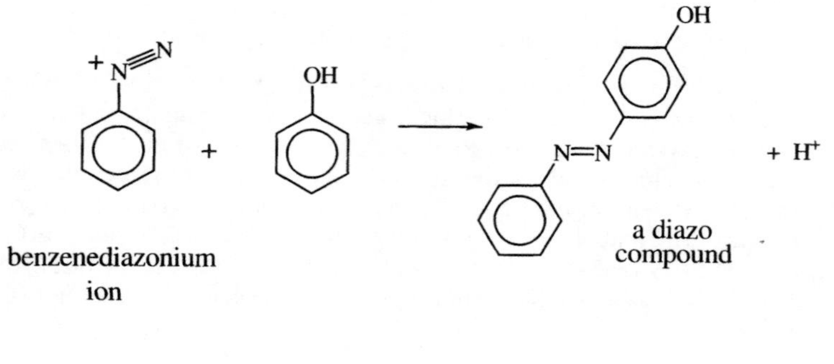

[Reaction scheme: benzenediazonium ion + phenol → a diazo compound (4-hydroxyphenyl-azo-benzene) + H⁺]

The coupled products are highly coloured and diazo compounds have been used extensively as dyes for dying wool and fabrics. A variety of different colours can be produced by changing the substituents on the aromatic ring of the aromatic diazonium salt or varying the nature of the phenol or aromatic amine which the diazonium ion couples with. The formation of diazo dyes is used as a qualitative test for the presence of a primary aromatic amine functional group in an organic compound. If a compound contains a primary aromatic amine functional group, then reaction with nitrous acid followed by a phenol will result in formation of a brightly coloured precipitate.

Amines Around Us

Most pharmaceuticals and natural products with therapeutic properties are amines. *Morphine* and *codeine* are examples of the plant-derived class of natural amines called *alkaloids*. Morphine is still employed as one of the most powerful pain-relieving compounds and codeine is used medicinally to treat a variety of conditions. *Caffeine* is a stimulant and is a component of coffee and tea. *Ephedrine* is used in the treatment of asthma and other bronchial disorders. *Histamine* is a compound released by the body when there is injury or irritation (the antihistamine drugs combat the inflammatory response casused by the release of histamine).

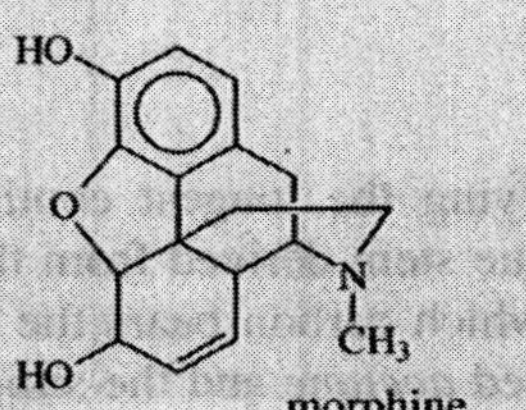

morphine

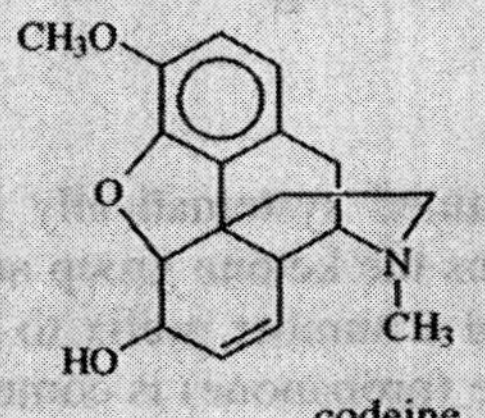

codeine

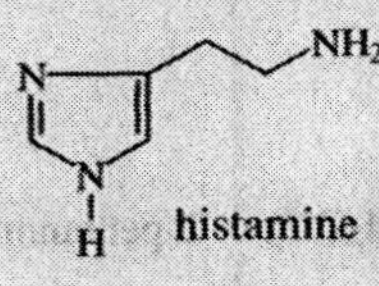

histamine

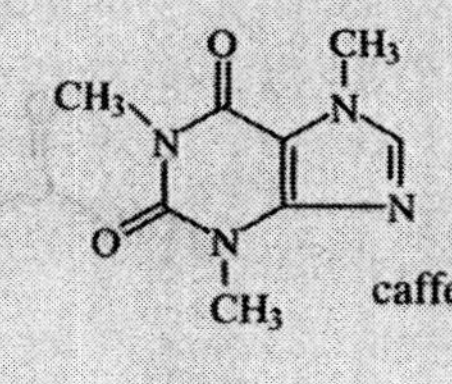

caffeine

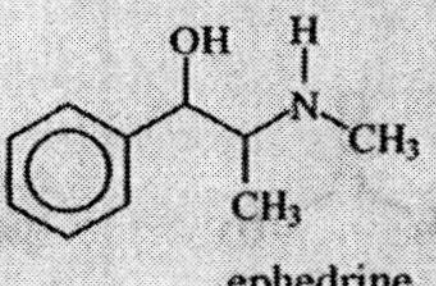

ephedrine

2.6 ALDEHYDES AND KETONES

Aldehydes and **ketones** contain the C=O (**carbonyl**) functional group. Aldehydes have at least one hydrogen attached to the carbon atom of the carbonyl group and ketones have two alkyl or aryl groups attached to the carbon atom of the carbonyl group.

Naming carbonyl compounds

Aldehydes are named systematically by identifying the longest continuous carbon chain which contains the aldehyde group and using the stem derived from this chain with the ending *-al*. The simplest aromatic aldehyde is called *benzaldehyde* and additionally, the lower aldehydes (C1—C4) all have non-systematic names which are in common use:

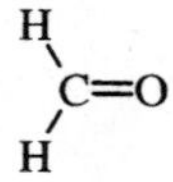

$$H_2C=O \quad \text{formaldehyde (methanal)}$$

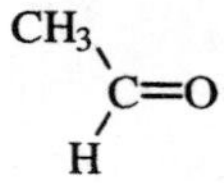

$$CH_3-CH=O \quad \text{acetaldehyde (ethanal)}$$

$$CH_3CH_2-CH=O \quad \text{propionaldehyde (propanal)}$$

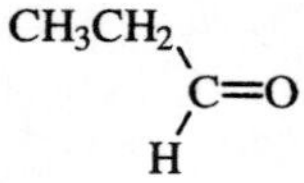

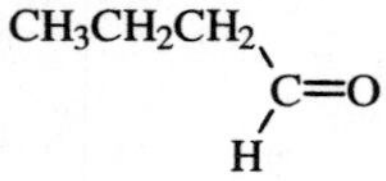

$$CH_3CH_2CH_2-CH=O \quad \text{butyraldehyde (butanal)}$$

Ketones are named systematically by identifying the longest continuous carbon chain which contains the ketone group and using the stem derived from this chain with the ending *-one* and a number prefix to indicate which carbon bears the oxygen atom. The simplest ketone (propanone) is commonly called *acetone* and the simplest aromatic ketone is called *acetophenone*.

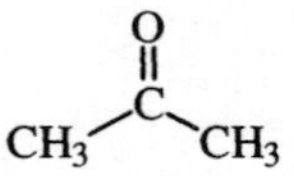

$CH_3-CO-CH_3$ acetone (propanone) $CH_3-CO-CH_2CH_2CH_3$ 2-pentanone

$CH_3-CO-CH_2CH_3$ butanone $CH_3CH_2-CO-CH_2CH_3$ 3-pentanone

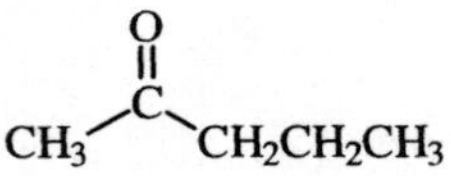

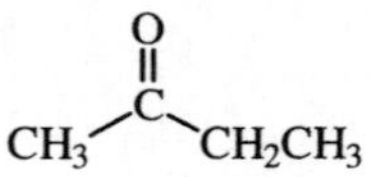

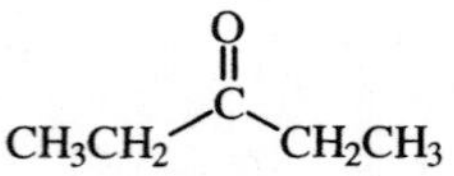

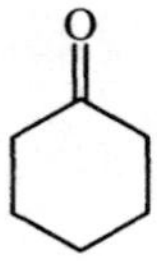

cyclohexanone $Ph-CO-CH_3$ acetophenone

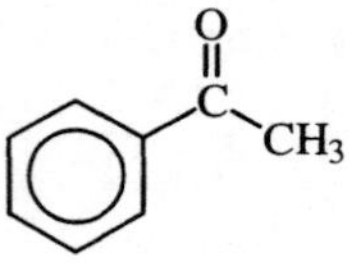

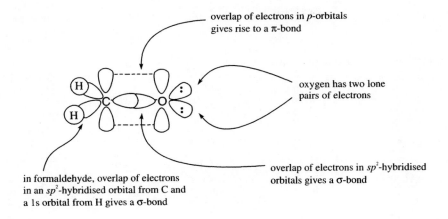

Figure 2.3 Bonding of the carbonyl group in formaldehyde

Bonding in aldehydes and ketones

The bonding in the carbonyl group is similar to that in an alkene (Figure 2.3 and Section 1.2). Both the carbon atom and the oxygen atom are sp^2-hybridised and, like the C=C bond of an alkene, the double bond is composed of a σ-bond and a π-bond. The oxygen atom of the carbonyl group has two non-bonding pairs of electrons occupying sp^2-hybridised orbitals. The carbonyl group is a highly polarised group with the oxygen atom being the more electronegative (δ−) end of the bond and the carbon the more electropositive (δ+). The polarity of the carbonyl group is particularly important since much of the reaction chemistry of aldehydes and ketones is dominated by the attack of nucleophiles at the carbon atom of the carbonyl group and the attack of electrophiles at the oxygen.

Nucleophilic addition reactions of aldehydes and ketones

Nucleophiles add to the carbon atom of the carbonyl group of aldehydes and ketones. The carbonyl C=O is a relatively polar bond and a great variety of nucleophiles react with this group. Attack of a nucleophile on a carbonyl group is accompanied by: (a) the formation of a new bond between the nucleophile and the carbonyl carbon, (b) a change in the hybridisation of the carbon from sp^2 to sp^3, and (c) a change in the geometry of the bonds about the carbon from trigonal planar to tetrahedral. When a nucleophile attacks a carbonyl group, the π-bond of the C=O is broken and the oxygen atom formally attains a negative charge, forming an alkoxide (Section 2.2). Reaction of the alkoxide with an acid (for example, water) affords an alcohol.

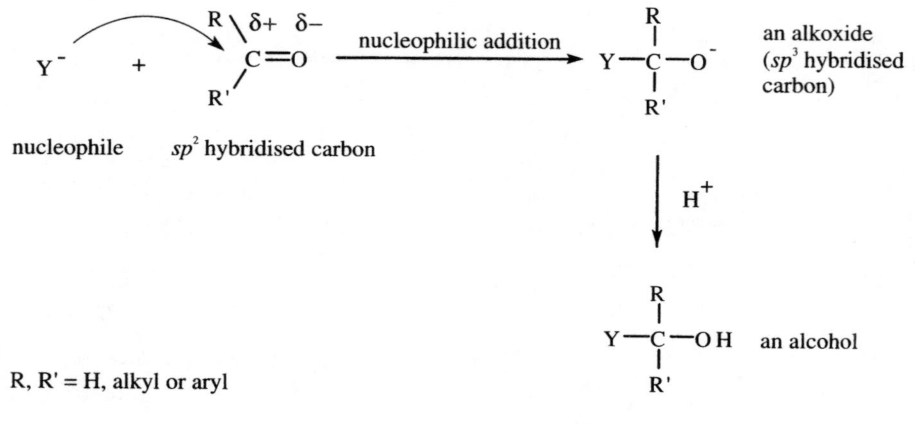

$$Y^- + \underset{\text{nucleophile}}{\underset{R'}{\overset{R}{\diagdown}}\!\!\!\!\!\!\overset{\delta+\ \delta-}{C=O}} \xrightarrow{\text{nucleophilic addition}} Y-\underset{R'}{\overset{R}{\underset{|}{\overset{|}{C}}}}-O^- \quad \begin{array}{l}\text{an alkoxide}\\(sp^3 \text{ hybridised}\\\text{carbon})\end{array}$$

nucleophile sp^2 hybridised carbon

$\downarrow H^+$

$$Y-\underset{R'}{\overset{R}{\underset{|}{\overset{|}{C}}}}-OH \quad \text{an alcohol}$$

R, R' = H, alkyl or aryl

Addition of Grignard reagents to aldehydes and ketones

Grignard reagents (organomagnesium reagents, Section 2.1) and organolithium reagents are excellent nucleophiles and readily undergo nucleophilic addition reactions with aldehydes and ketones. Both Grignard reagents and organolithium reagents are effectively sources of alkyl or aryl carbanions (R^-) and the attack of these reagents on carbonyl compounds results in the formation of a new C—C bond. The addition of a Grignard reagent to formaldehyde results in the formation of a primary alcohol after the intermediate alkoxide salt is added to water or another source of H^+. Note that this is a general method for extending the carbon chain of a molecule by one carbon atom.

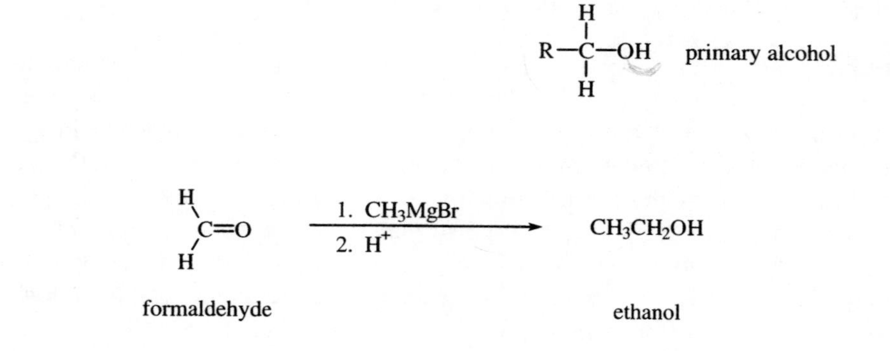

$$\underset{\substack{\text{Grignard}\\\text{reagent}}}{RMgBr} + \underset{H}{\overset{H}{\diagdown}}\!\!\!\!\!\!\overset{\delta+\ \delta-}{C=O} \longrightarrow R-\underset{H}{\overset{H}{\underset{|}{\overset{|}{C}}}}-O^-\ MgBr^+ \quad \begin{array}{l}\text{an alkoxide salt}\\\text{of a primary alcohol}\end{array}$$

$\downarrow H^+$

$$R-\underset{H}{\overset{H}{\underset{|}{\overset{|}{C}}}}-OH \quad \text{primary alcohol}$$

$$\underset{\text{formaldehyde}}{\underset{H}{\overset{H}{\diagdown}}\!\!\!\!\!\!C=O} \xrightarrow[\text{2. } H^+]{\text{1. } CH_3MgBr} \underset{\text{ethanol}}{CH_3CH_2OH}$$

The addition of a Grignard reagent to any other aldehyde results in the formation of a secondary alcohol after the intermediate alkoxide salt is added to water or another source of H^+. This is a general method of making secondary alcohols. By choosing the appropriate aldehyde and Grignard reagent it is possible to synthesise many secondary alcohols.

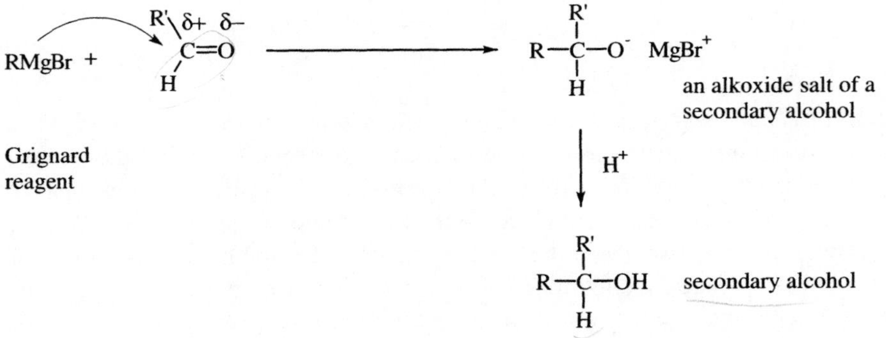

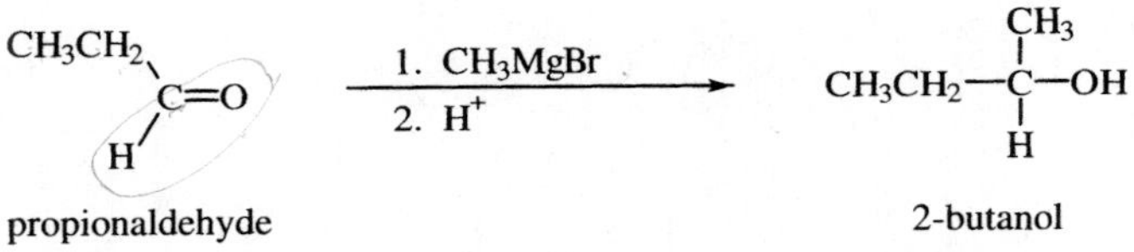

$$RMgBr + \underset{H}{\overset{R'}{>}}C=O \longrightarrow R-\underset{H}{\overset{R'}{\underset{|}{C}}}-O^- \; MgBr^+ \quad \text{an alkoxide salt of a secondary alcohol}$$

Grignard reagent

$$\downarrow H^+$$

$$R-\underset{H}{\overset{R'}{\underset{|}{C}}}-OH \quad \text{secondary alcohol}$$

$$\underset{H}{\overset{CH_3CH_2}{>}}C=O \quad \xrightarrow[\text{2. } H^+]{\text{1. } CH_3MgBr} \quad CH_3CH_2-\underset{H}{\overset{CH_3}{\underset{|}{C}}}-OH$$

propionaldehyde — 2-butanol

The addition of a Grignard reagent to a ketone results in the formation of a tertiary alcohol after the intermediate alkoxide salt is added to water or another source of H^+. Again, this is a general reaction for the synthesis of tertiary alcohols.

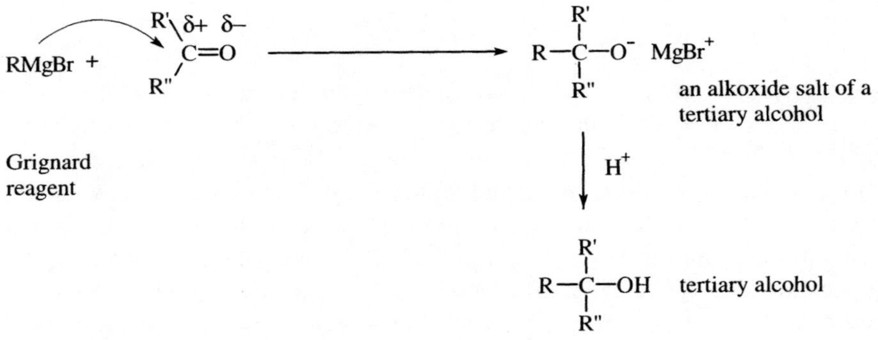

$$RMgBr + \underset{R''}{\overset{R'}{>}}C=O \longrightarrow R-\underset{R''}{\overset{R'}{\underset{|}{C}}}-O^- \; MgBr^+ \quad \text{an alkoxide salt of a tertiary alcohol}$$

Grignard reagent

$$\downarrow H^+$$

$$R-\underset{R''}{\overset{R'}{\underset{|}{C}}}-OH \quad \text{tertiary alcohol}$$

Derivatives of Hydrocarbons **93**

$$\underset{\text{butanone}}{\underset{CH_3}{\overset{CH_3CH_2}{>}}C=O} \quad \xrightarrow[\text{2. H}^+]{\text{1. CH}_3\text{MgBr}} \quad \underset{\text{2-methyl-2-butanol}}{CH_3CH_2-\underset{\underset{CH_3}{|}}{\overset{\overset{CH_3}{|}}{C}}-OH}$$

Addition of carbon nucleophiles to aldehydes and ketones—the aldol condensation
Hydrogen atoms attached to carbon atoms adjacent to the carbonyl group of an aldehyde or ketone are more acidic than most alkyl protons and can be removed under mildly basic conditions. The carbon atom adjacent to a functional group in an alkyl chain is frequently termed the α-*carbon*, the next carbon is called the β-*carbon*, and so on. The acidity of the hydrogen atoms attached to the α-carbon of the carbonyl group arises because the conjugate base formed on deprotonation is unusually stable. The anion is stabilised by delocalisation of the negative charge onto both the α-carbon and the oxygen atom of the carbonyl group and as a consequence, the hydrogen atoms adjacent to the C=O are relatively easy to remove. The resulting anion is termed an *enolate anion* and is a good nucleophile.

$$R-CH_2-\overset{\overset{O}{\|}}{C}-R' \quad \xrightarrow{\text{base}} \quad \left[\underset{H \quad R'}{\overset{R \quad O^-}{>}}C=C< \quad \longleftrightarrow \quad \underset{H \quad R'}{\overset{R \quad O}{>}}\overset{-}{C}-\overset{\|}{C}< \right]$$

hydrogen atoms at
the α-carbon are acidic

resonance stabilised enolate ion

When an aldehyde or ketone with hydrogen atoms attached to the α-carbon atom is treated with a base, an enolate ion is formed, at least in low concentration. The nucleophilic addition of the enolate to the C=O of another molecule leads to the formation of a new C—C bond and an alkoxide. The alkoxide is protonated by the reaction solvent or by the acidic protons adjacent to the C=O of another molecule of the aldehyde or ketone, to yield a β-hydroxyaldehyde or β-hydroxyketone. The overall reaction of two molecules of an aldehyde or ketone in the presence of a base to form a β-hydroxyaldehyde or β-hydroxyketone is called the *aldol condensation*.

$$R'-\overset{\underset{\parallel}{O}}{C}-CH_2-R \underset{acid}{\overset{base}{\rightleftharpoons}} R'-\overset{\underset{\parallel}{O}}{C}-\overset{-}{C}H-R$$

$$R'-\overset{\underset{\parallel}{O}}{C}-CH_2-R + R'-\overset{\underset{\parallel}{O}}{C}-\overset{-}{C}H-R \rightleftharpoons R'-\underset{\underset{R'-\overset{\underset{\parallel}{O}}{C}-CH-R}{|}}{\overset{\underset{|}{O^-}}{C}}-CH_2-R \xrightarrow{H^+} R'-\overset{\underset{\parallel}{O}}{C}-\underset{\underset{R}{|}}{C}H-\underset{\underset{R'}{|}}{\overset{\underset{|}{OH}}{C}}-CH_2R$$

aldol condensation   a β-hydroxyketone

The β-hydroxyaldehydes or β-hydroxyketones formed in the aldol condensation are more easily dehydrated than simple alcohols. Quite often no additional reagents are required to cause the dehydration, and the reaction conditions which give rise to the aldol condensation are sufficient to cause dehydration of the aldol product to give a compound containing both a C=O and a C=C functional group. Even though more than one dehydration product can be possible, the product formed invariably has the C=O and the C=C groups in a conjugated arrangement (an α, β-unsaturated aldehyde or ketone).

$$R'-\overset{\underset{\parallel}{O}}{C}-\underset{\underset{H\ R}{|}}{\overset{\underset{|}{R'}}{C}}_\alpha-\underset{\beta}{\overset{\underset{|}{OH}}{C}}-\underset{\gamma}{C}R \xrightarrow{dehydration} R'-\overset{\underset{\parallel}{O}}{C}-\underset{\underset{R}{}}{\overset{\underset{}{R'}}{C}}_\alpha=\underset{\beta}{C}-R$$

an α,β-unsaturated
ketone or aldehyde

SAMPLE EXERCISE 2.12

Give the condensed structural formula of the β-hydroxyaldehyde product which arises when propanal is treated with sodium hydroxide.

Solution: The enolate arising from deprotonation at the carbon atom α to the C=O group of propanal is a good nucleophile and attacks the carbonyl group of another molecule of propanal. The resulting alkoxide ion is protonated by a proton source in solution to give 3-hydroxy-2-methylpentanal (a β-hydroxyaldehyde).

Derivatives of Hydrocarbons **95**

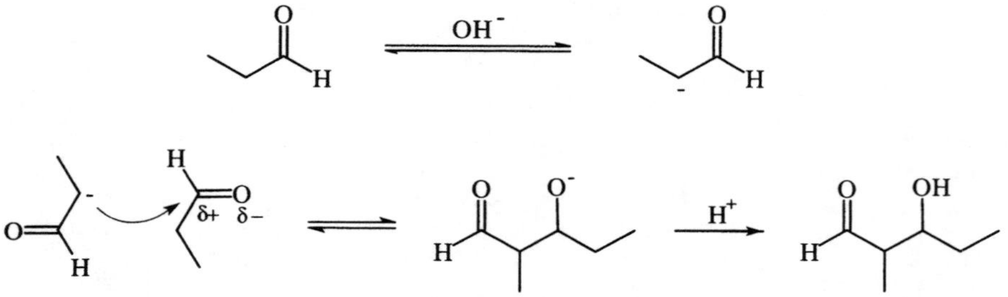

3-hydroxy-2-methylpentanal

PRACTICE EXERCISE

Give the condensed structural formulas for the α,β-unsaturated aldehydes or ketones which result from dehydration of the β-hydroxyaldehydes or β-hydroxyketones formed by aldol condensation of the following compounds:

(a) 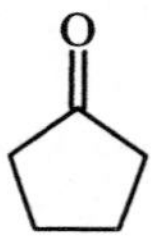   cyclopentanone   (b) 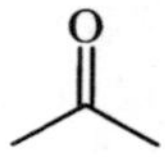   acetone

Answers:

(a) 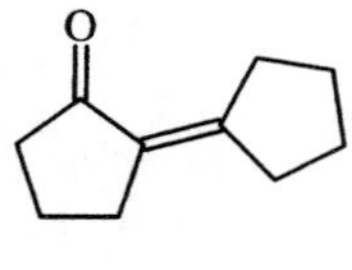 cyclopentylidenecyclopentanone   (b) 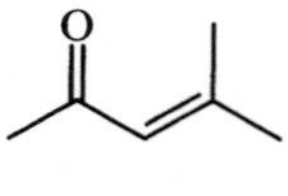 4-methylpent-3-en-2-one

Reduction of aldehydes and ketones with hydride reducing agents

Aldehydes and ketones are reduced to alcohols by reagents which act as a source of nucleophilic H⁻ (hydride ions). Reagents such as sodium borohydride (NaBH$_4$) and lithium aluminium hydride (LiAlH$_4$) act as if they are sources of H⁻ nucleophiles when they react with aldehydes and ketones. Aldehydes are reduced by NaBH$_4$ or LiAlH$_4$ by addition of hydride to the carbon of the carbonyl group. After reduction and hydrolysis of the intermediate alkoxide salts, aldehydes give rise to primary alcohols and ketones give rise to secondary alcohols.

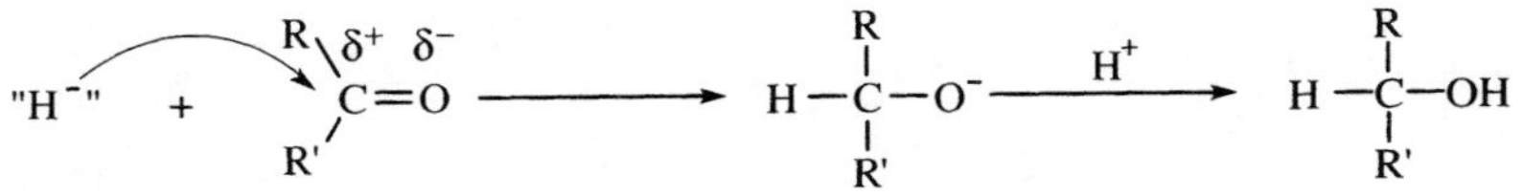

$$\text{"H}^-\text{"} + \underset{R'}{\overset{R}{\diagdown}}\overset{\delta^+\ \delta^-}{C=O} \longrightarrow H-\underset{R'}{\overset{R}{\underset{|}{C}}}-O^- \xrightarrow{H^+} H-\underset{R'}{\overset{R}{\underset{|}{C}}}-OH$$

Each mole of LiAlH$_4$ or NaBH$_4$ effectively produces four equivalents of hydride and can reduce four moles of aldehyde or ketone. In practice LiAlH$_4$ reacts violently with water or alcohols to produce hydrogen gas and reductions with LiAlH$_4$ must be carried out carefully in solvents which have been scrupulously dried to remove all traces of moisture. NaBH$_4$ is less reactive than LiAlH$_4$ and reductions can be carried out in methanol or methanol/water mixtures.

SAMPLE EXERCISE 2.13

Give the structural formula for the organic product which is formed when cyclopentanone is treated with lithium aluminium hydride (in dry diethyl ether solvent) and then treated with a dilute solution of HCl.

Solution: Cyclopentanone is a ketone and will be reduced to a secondary alcohol. The LiAlH$_4$ reagent is effectively a source of H$^-$ ions which add as a nucleophile to the C=O group of the ketone to form an alkoxide salt. The addition of dilute acid converts the alkoxide to the free alcohol.

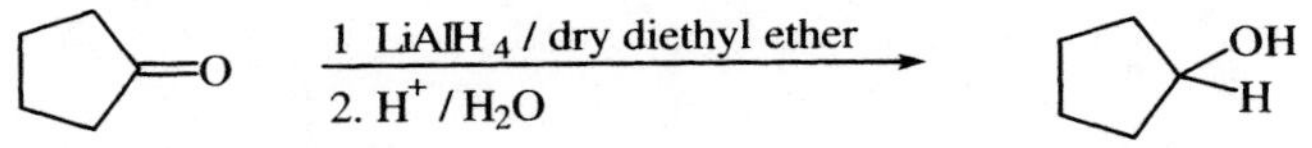

cyclopentanone $\xrightarrow[\text{2. H}^+/\text{H}_2\text{O}]{\text{1. LiAlH}_4/\text{dry diethyl ether}}$ cyclopentanol

PRACTICE EXERCISE

Give the structural formulas for the organic products formed when each of the following compounds is treated with sodium borohydride (in methanol solvent) and then is treated with a dilute aqueous solution of HCl.

(a) 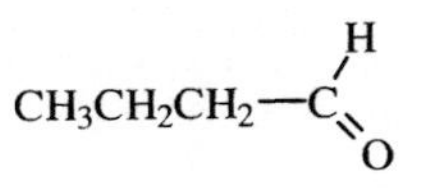 $CH_3CH_2CH_2-\overset{H}{\underset{O}{C}}$

(b) $CH_3CH_2-\overset{CH_3}{\underset{O}{C}}$

(c) $CH_3-\underset{\overset{|}{\underset{O}{C}}{\underset{H}{}}}{CH}-CH_3$

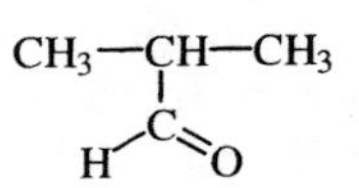

(d) 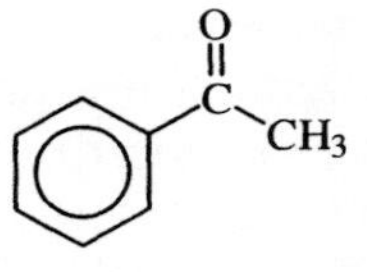 phenyl-C(=O)-CH$_3$

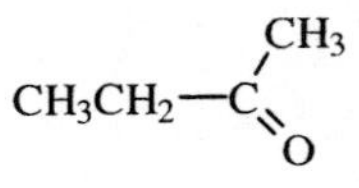

Answers:

(a) CH₃CH₂CH₂CH₂OH

(b) 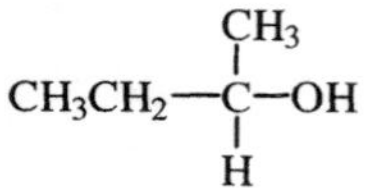 CH₃CH₂—C(CH₃)(H)—OH

(c) CH₃—CH(CH₂OH)—CH₃

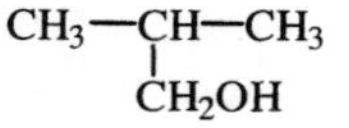

(d) 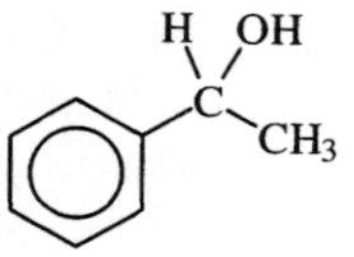 C₆H₅—C(H)(OH)(CH₃)

Addition of nitrogen nucleophiles to aldehydes and ketones

The nucleophilic addition of ammonia or primary amines to aldehydes and ketones gives rise to compounds containing the **imine** (—C=N—) functional group. The products formed initially are amino-alcohols but these are unstable and lose H₂O rapidly to give imines (also known as *Schiff's bases*).

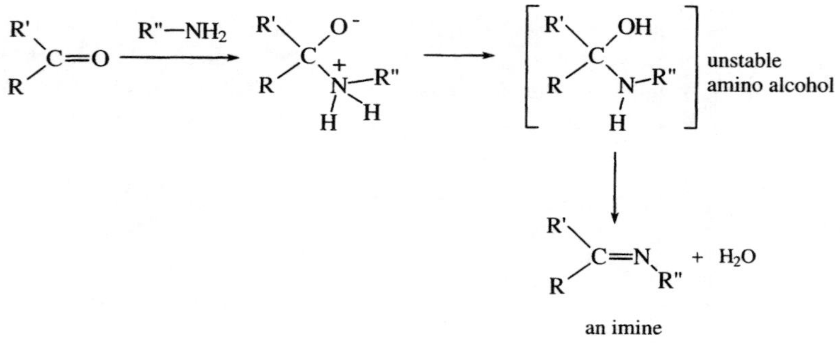

$$R'R C=O \xrightarrow{R''-NH_2} \underset{\underset{H\;H}{\overset{|}{N^+}-R''}}{\overset{O^-}{\underset{|}{C}}} \longrightarrow \left[\underset{\underset{H}{\overset{|}{N}-R''}}{\overset{OH}{\underset{|}{C}}} \right] \text{ unstable amino alcohol}$$

$$\downarrow$$

$$\underset{R}{\overset{R'}{>}}C=N_{R''} + H_2O$$

an imine

The reaction which forms an imine can be reversed by treating the imine with water, typically in a dilute acid solution. This reaction is called *a hydrolysis reaction* and liberates the amine (in its protonated form in acidic solution) and the ketone or aldehyde.

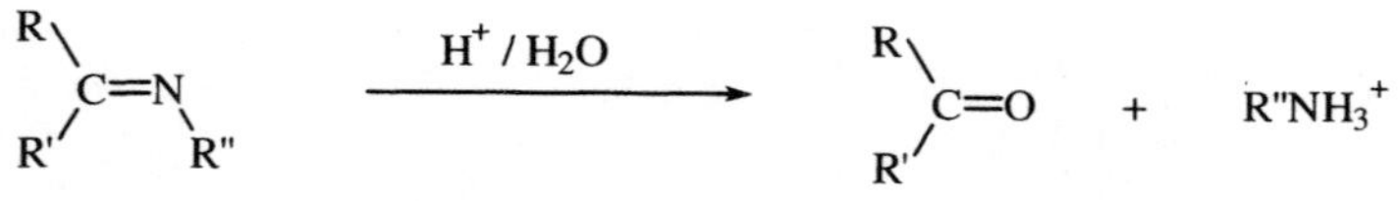

$$\underset{R'}{\overset{R}{>}}C=N_{R''} \xrightarrow{H^+/H_2O} \underset{R'}{\overset{R}{>}}C=O + R''NH_3^+$$

Other compounds containing the amino group also act as good nucleophiles towards the carbonyl group of aldehydes and ketones. In particular hydroxylamine, hydrazine, 2,4-dinitrophenylhydrazine and semicarbazide all contain the amino functional group and all give products analogous to imines when they react with aldehydes and ketones. The products of reaction of an aldehyde or ketone with hydroxylamine are called *oximes*; the products obtained on reaction with semicarbazide are called *semicarbazones* and the products obtained on reaction with hydrazine are called *hydrazones*, thus the products obtained on reaction with 2,4-dinitrophenylhydrazine are *2,4-dinitrophenyhydrazones*. The formation of oximes, hydrazones, 2,4-dinitrophenylhydrazones and semicarbazones is analogous to the formation of imines, only the —NH_2 part of the reagents is involved in the reaction.

$$\underset{\substack{R, R' = H, \text{ alkyl} \\ \text{or aryl}}}{\overset{R}{\underset{R'}{>}}\!\!C=O} + \underset{\text{hydroxylamine}}{H_2N\!-\!OH} \longrightarrow \underset{\text{an oxime derivative}}{\overset{R}{\underset{R'}{>}}\!\!C=N\!-\!OH}$$

$$\overset{R}{\underset{R'}{>}}\!\!C=O + \underset{\text{hydrazine}}{H_2N\!-\!NH_2} \longrightarrow \underset{\text{a hydrazone derivative}}{\overset{R}{\underset{R'}{>}}\!\!C=N\!-\!NH_2}$$

$$\overset{R}{\underset{R'}{>}}\!\!C=O + \underset{\text{2,4-dinitrophenylhydrazine}}{H_2N\!-\!NH\!-\!C_6H_3(NO_2)_2} \longrightarrow \underset{\text{a 2,4-dinitrophenylhydrazone derivative}}{\overset{R}{\underset{R'}{>}}\!\!C=N\!-\!NH\!-\!C_6H_3(NO_2)_2}$$

$$\overset{R}{\underset{R'}{>}}\!\!C=O + \underset{\text{semicarbazide}}{H_2N\!-\!NHCNH_2\ (C=O)} \longrightarrow \underset{\text{a semicarbazone derivative}}{\overset{R}{\underset{R'}{>}}\!\!C=N\!-\!NHCNH_2\ (C=O)}$$

Hydroxylamine, hydrazine, 2,4-dinitrophenylhydrazine and semicarbazide are frequently used as reagents to form crystalline derivatives of aldehydes and ketones. These derivatives are usually easier to isolate and to purify than the original aldehyde or ketone and usually have sharp melting points. Most reference books of data on organic compounds contain the characteristic melting points of the oxime, hydrazone, 2,4-dinitrophenylhydrazone and semicarbazone derivatives of known aldehydes and ketones. 2,4-dinitrophenylhydrazone derivatives are typically bright orange or yellow solids and the formation of an orange or yellow precipitate when an unknown compound is added to an acidified solution of 2,4-dinitrophenylhydrazine is used as a qualitative test to indicate the presence of an aldehyde or ketone functional group in the molecule.

Addition of oxygen nucleophiles to aldehydes and ketones
The oxygen atoms of both water and alcohols add as nucleophiles to the C=O group of aldehydes and ketones. The addition of water forms compounds which contain two —OH groups attached to the same carbon atom (called a *1,1-diol*, a *geminal diol* or a *hydrate*). When an aldehyde or ketone is added to water, an equilibrium mixture of the compound with its hydrate is formed:

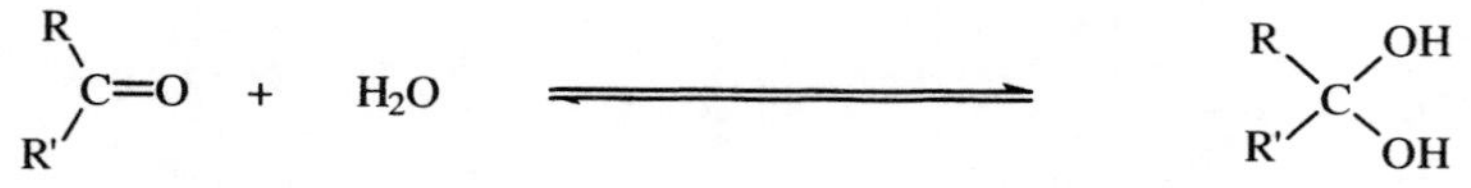

$$\underset{R,\ R'\ =\ H,\ alkyl\ or\ aryl}{\underset{R'}{\overset{R}{>}}C=O} + H_2O \rightleftharpoons \underset{a\ hydrate}{\underset{R'}{\overset{R}{>}}C\underset{OH}{\overset{OH}{<}}}$$

The position of the equilibrium, that is, whether the free aldehyde or ketone is the major species in solution or whether the hydrate is the major species, depends on the specific aldehyde or ketone being considered. For formaldehyde, the equilibrium lies well on the side of the hydrate—free formaldehyde which forms less than 1 per cent of a mixture of formaldehyde and water. For other aldehydes, the amounts of free aldehyde and hydrate in a mixture of water and aldehyde are similar. For ketones, typically the equilibrium strongly favours the free ketone.

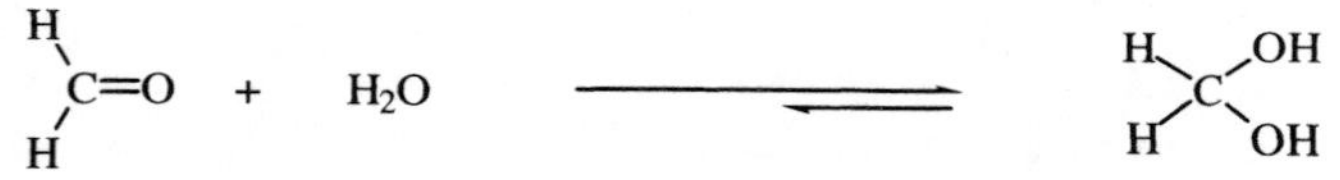

$$\underset{\substack{formaldehyde \\ <1\%}}{\underset{H}{\overset{H}{>}}C=O} + H_2O \rightleftharpoons \underset{\substack{formaldehyde\ hydrate \\ <99\%}}{\underset{H}{\overset{H}{>}}C\underset{OH}{\overset{OH}{<}}}$$

$$\underset{\substack{comparable\ concentrations \\ of\ aldehyde\ and\ hydrate}}{\underset{H}{\overset{R}{>}}C=O} + H_2O \rightleftharpoons \underset{an\ aldehyde\ hydrate}{\underset{H}{\overset{R}{>}}C\underset{OH}{\overset{OH}{<}}}$$

$$\underset{ketone\ dominates}{\underset{R'}{\overset{R}{>}}C=O} + H_2O \rightleftharpoons \underset{ketone\ hydrate}{\underset{R'}{\overset{R}{>}}C\underset{OH}{\overset{OH}{<}}}$$

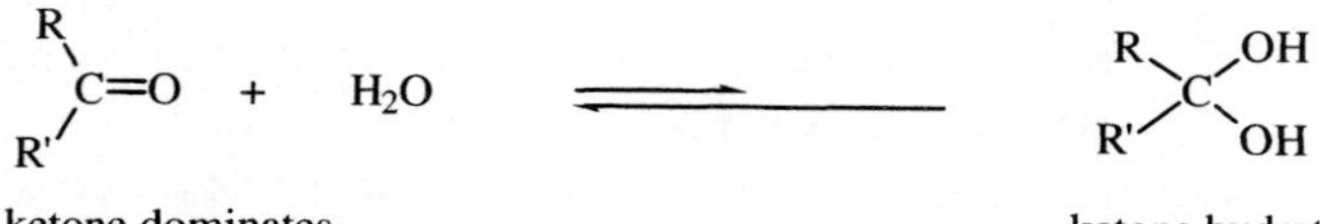

Alcohols add to aldehydes and to ketones to form **hemiacetals**. The reaction is analogous to the addition of water—when an aldehyde or ketone is dissolved in an alcohol solution, the free carbonyl compound is in equilibrium with its hemiacetal.

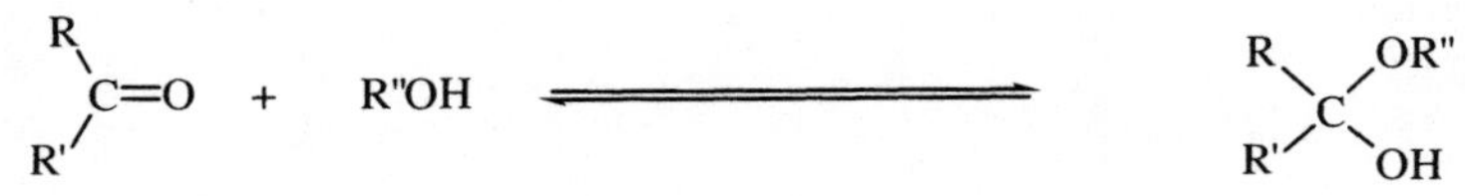

$$\underset{R'}{\overset{R}{>}}C=O \;+\; R''OH \;\rightleftharpoons\; \underset{R'}{\overset{R}{>}}C\underset{OH}{\overset{OR''}{<}}$$

R, R' = H, alkyl or aryl   a hemiacetal

Typically the equilibrium strongly favours the free aldehyde or ketone *except* when an aldehyde or ketone can react intramolecularly (that is, within the same molecule), particularly if a five- or six-membered ring is formed, then the cyclic hemiacetal is the major species in the equilibrium.

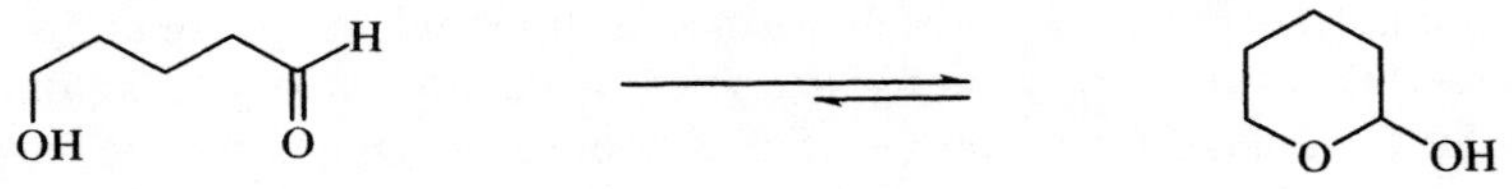

5-hydroxypentanal   a cyclic hemiacetal
(minor component)   (major component)

Simple sugars (including glucose) are polyhydroxyaldehydes and they exist in solution as an equilibrum mixture of open chain hydroxy-aldehyde and cyclic hemiacetal compounds.

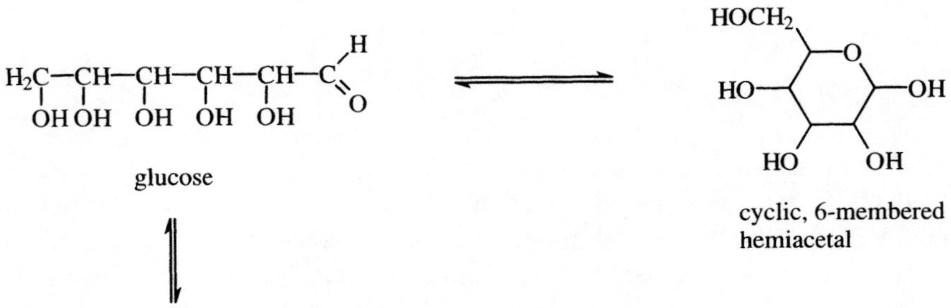

glucose ⇌ cyclic, 6-membered hemiacetal

↕

other cyclic hemiacetals

Hemiacetals react with alcohols in the presence of an acid catalyst to form **acetals**.

Derivatives of Hydrocarbons   **101**

$$\underset{\text{a hemiacetal}}{\overset{R}{\underset{R'}{>}}C\overset{OR''}{\underset{OH}{<}}} + R''OH \underset{}{\overset{H^+ \text{ catalyst}}{\rightleftharpoons}} \underset{\text{an acetal}}{\overset{R}{\underset{R'}{>}}C\overset{OR''}{\underset{OR''}{<}}} + H_2O$$

The reaction is reversible (an equilibrium) and to ensure that high yields of the acetals are obtained, it is usually necessary to remove water from the reaction mixture by distillation or using a water-absorbing material. Since the reaction is reversible, the ketone or aldehyde can be reformed from the acetal by treatment with water, usually in the presence of an acid catalyst (a hydrolysis reaction). This means that formation of an acetal can be used to 'protect' an aldehyde or ketone temporarily while some reaction chemistry is performed on functional groups elsewhere in the molecule, then the aldehyde or ketone can be reformed by treatment with dilute acid.

If a diol is used to form the hemiacetals, then the second —OH group can react to form a cyclic acetal. The cyclic acetals formed using ethylene glycol (1,2-ethanediol) are frequently used as temporary *protecting groups* for aldehydes and ketones.

In a similar fashion, ketones can be used as protecting groups for diols. A diol can be conveniently converted to an acetal by reaction with a simple aldehyde or ketone and when desired, the diol can be unmasked by treatment with dilute acid.

Oxidation of aldehydes

The oxidation of aldehydes is a facile process, as was indicated when the oxidation of primary alcohols was considered (Section 2.2). Aldehydes are easily oxidised to carboxylic acids by the normal laboratory oxidising agents ($Cr_2O_7^{2-}$ in acidic solution, MnO_4^- in acidic or basic solution, CrO_3 in acidic solution) as well as a variety of other reagents, such as oxygen (air) and silver ions in basic solution. **Tollens reagent** is a solution of $[Ag(NH_3)_2]^+$ in dilute sodium hydroxide solution and this is used as a qualitative test for the presence of the aldehyde functional group. Tollens reagent oxidises the aldehyde functional group and in the process Ag^+ is reduced to metallic silver which is deposited as a silver mirror on the reaction vessel.

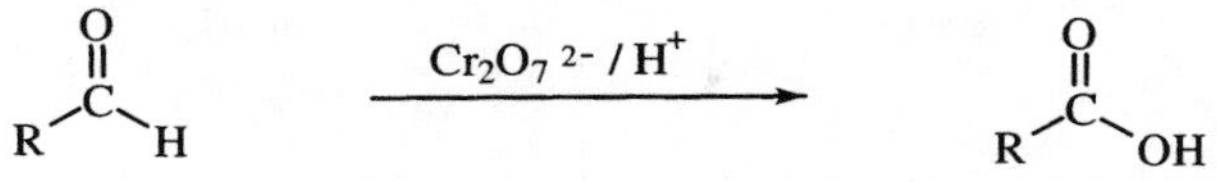

$$\underset{R}{\overset{O}{\underset{H}{\|}}}\!\!C \quad \xrightarrow{Cr_2O_7{}^{2-}/H^+} \quad \underset{R}{\overset{O}{\underset{OH}{\|}}}\!\!C$$

Catalytic reduction of aldehydes and ketones

The C=O group of aldehydes and ketones is reduced by hydrogen gas when a powdered metal catalyst of palladium, platinum or nickel is used, analagous to the reduction of alkenes. In general the carbonyl group is not as readily reduced as the C=C group of alkenes and elevated temperatures and pressures are usually used. Aldehydes give rise to primary alcohols on reduction with hydrogen over a metal catalyst and ketones give rise to secondary alcohols. Unlike the reduction of aldehydes and ketones with metal hydrides (which also form primary and secondary alcohols), the reduction using hydrogen can be complicated by the fact that other functional groups, such as alkenes and alkynes, present in the molecule can also be reduced.

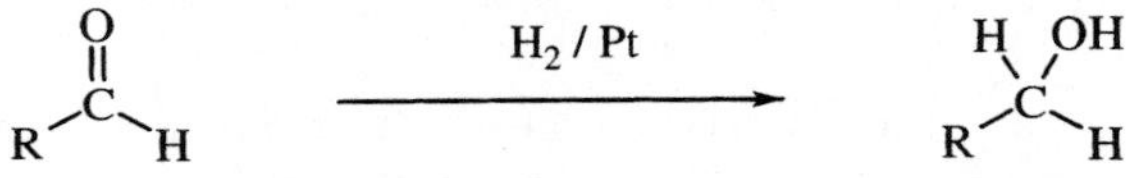

$$\underset{\text{aldehyde}}{R-\overset{O}{\underset{\|}{C}}-H} \quad \xrightarrow{H_2/Pt} \quad \underset{\text{primary alcohol}}{R-\overset{H}{\underset{H}{C}}-OH}$$

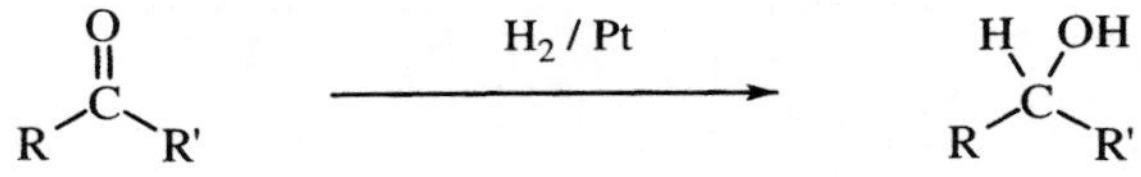

$$\underset{\text{ketone}}{R-\overset{O}{\underset{\|}{C}}-R'} \quad \xrightarrow{H_2/Pt} \quad \underset{\text{secondary alcohol}}{R-\overset{H}{\underset{R'}{C}}-OH}$$

Aldehydes and Ketones Around Us

Cinnamaldehyde is a fragrant oil responsible for the scent and taste of cinnamon spice. *Citronellal* is a natural oil found in the leaves of many citrus plants and has the tangy odour of citrus fruits. *Progesterone* is one of the principal female sex hormones; it is a steroid which is important for preparing the female uterus for implantation of a fertilised egg during pregnancy. *Camphor* is a natural ketone derived from camphor laurel trees and it is used medicinally.

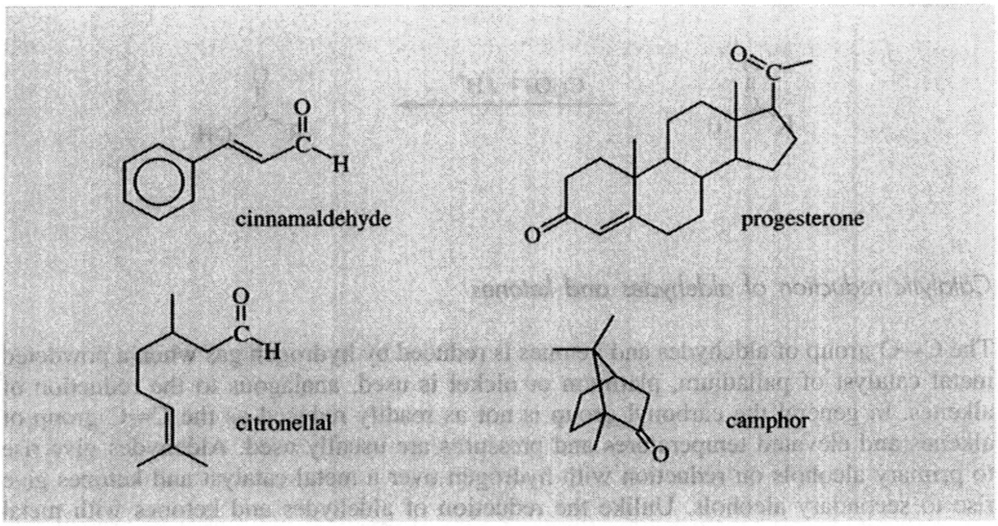

cinnamaldehyde

progesterone

citronellal

camphor

2.7 CARBOXYLIC ACIDS AND CARBOXYLIC ACID DERIVATIVES

Carboxylic acids contain the —COOH or carboxyl functional group. The carboxylic acid group is the most acidic of the common organic functional groups although its acid strength is still weak compared to the strength of common inorganic or mineral acids, (Table 2.4).

The acidity of carboxylic acids arises from the fact that the carboxylate ion (the conjugate base of the carboxylic acid) is a resonance stabilised carbanion with the negative charge delocalised over both oxygen atoms of the carboxylate group.

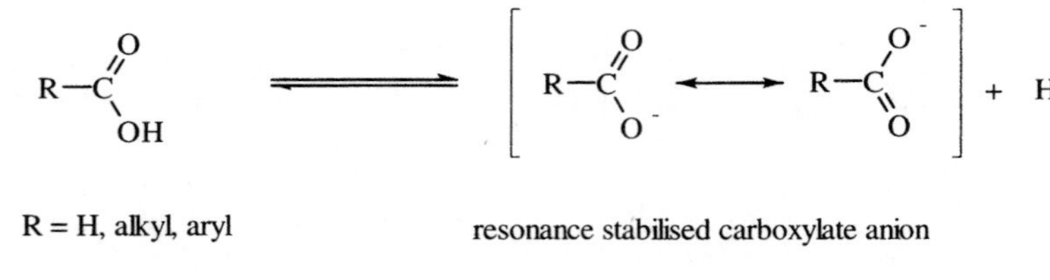

R = H, alkyl, aryl

resonance stabilised carboxylate anion

Naming carboxylic acids

Carboxylic acids are named by identifying the longest carbon chain containing the carboxylic acid functional group and using this carbon chain as the stem for the carboxylic acid name. The ending *-anoic acid* is added to the stem to indicate that the compound is a carboxylic acid. When a carboxylic acid is deprotonated, the resulting carboxylate ion is termed an alkanoate ion. The conjugate bases are named by replacing the *-ic* ending of the carboxylic acid name with *-ate*. The R—CO fragment derived from a carboxylic acid is called an *acyl* group and is named by replacing the *-ic* ending of the carboxylic acid name with *-yl*. Most of the smaller carboxylic acids have common names that are still widely used. The systematic names are given in parentheses in Table 2.5.

Table 2.4 Strengths of some common inorganic and organic acids

Acid	pK_a
HCl	< −2
HNO_3	−1.37
HNO_2	3.15
HCOOH	3.74
CH_3COOH	4.76
CH_3CH_2COOH	4.87
$CH_3(CH_2)_2COOH$	4.82

SAMPLE EXERCISE 2.14

Name the following compounds:

(a) $CH_3CH_2CH_2CH_2COOH$

(b) [structure: 2,2-dimethylbutanoic acid]

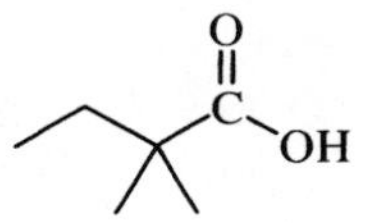

(c) [structure: E-2-pentenoic acid]

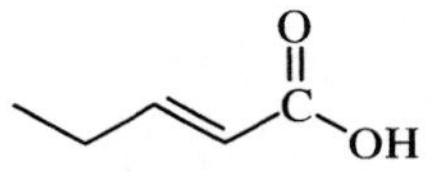

Solution: In (a) the longest chain of carbon atoms containing the carboxyl functional group has five carbon atoms. The stem for this chain is *pentan-* so the carboxylic acid is named *pentanoic acid*. In (b) the longest chain of carbon atoms containing the carboxyl functional group has four carbon atoms and two methyl substituents at carbon 2. The name is 2,2-dimethylbutyric acid (or 2,2-dimethylbutanoic acid). In (c) the longest chain of carbon atoms containing the carboxyl functional group has five carbon atoms and a double bond between carbon atoms 2 and 3 (numbering from the —COOH end of the chain). The stem for this chain is *penten-* so the carboxylic acid would be named 2-pentenoic acid. Note that with this compound, there is the possibility of having *E* or *Z* stereoisomers about the C=C and the isomer drawn has the *E* or *trans* stereochemistry. The full name of the compound is *E*-2-pentenoic acid.

Properties and reactions of carboxylic acids

Carboxylic acids are among the most polar of all organic compounds. The oxygen atoms of the carboxyl group are polarised strongly δ− and the hydrogen of the —OH group is strongly δ+. This means that, in a similar fashion to alcohol molecules, carboxylic acids can participate in hydrogen bonding. For carboxylic acids this is a particularly strong interaction and carboxylic acids frequently display molecular properties more characteristic of dimeric species than isolated molecules.

Table 2.5 Names of simple carboxylic acids

Carboxylic Acid	Carboxylate ion	Acyl group
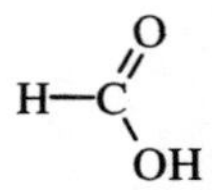 H—C(=O)OH formic acid (methanoic acid)	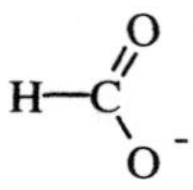 H—C(=O)O⁻ formate ion (methanoate ion)	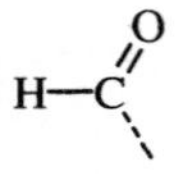 H—C(=O)— formyl group (methanoyl group)
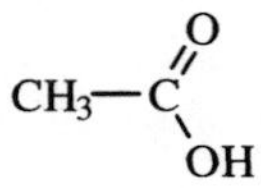 CH_3—C(=O)OH acetic acid (ethanoic acid)	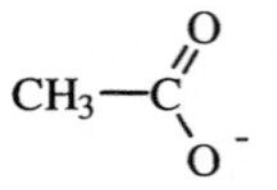 CH_3—C(=O)O⁻ acetate ion (ethanoate ion)	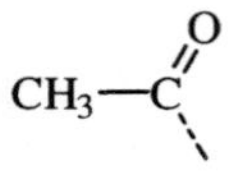 CH_3—C(=O)— acetyl group (ethanoyl group)
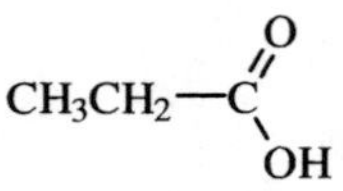 CH_3CH_2—C(=O)OH propionic acid (propanoic acid)	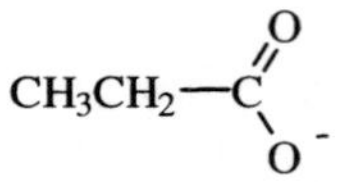 CH_3CH_2—C(=O)O⁻ propionate ion (propanoate ion)	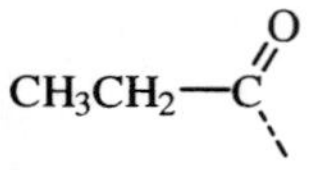 CH_3CH_2—C(=O)— propionyl group (propanoyl group)
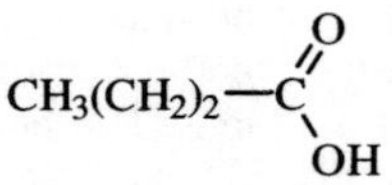 $CH_3(CH_2)_2$—C(=O)OH butyric acid (butanoic acid)	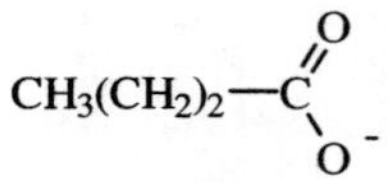 $CH_3(CH_2)_2$—C(=O)O⁻ butyrate ion (butanoate ion)	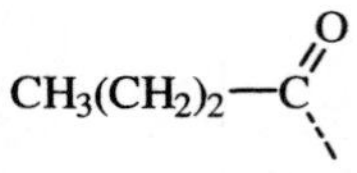 $CH_3(CH_2)_2$—C(=O)— butyryl group (butanoyl group)
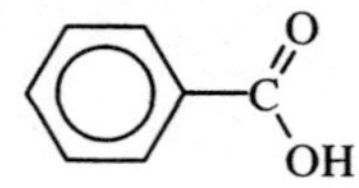 C_6H_5—C(=O)OH benzoic acid	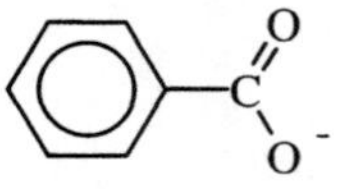 C_6H_5—C(=O)O⁻ benzoate ion	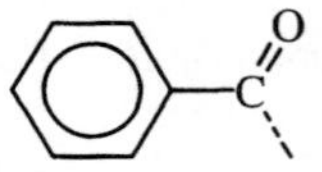 C_6H_5—C(=O)— benzoyl group

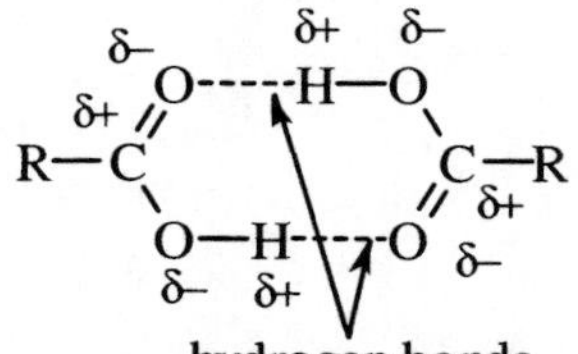

Carboxylic acids are sufficiently strong acids to react with hydroxide, carbonate, and hydrogencarbonate ions in aqueous solution. Carboxylic acids liberate CO_2 from solutions of CO_3^{2-} or HCO_3^- and this is frequently used as a simple qualitative test for the presence of the carboxylic acid group. The carboxylate ions are generally water-soluble and carboxylic acids can be dissolved in water by ensuring that the water is basic. Deprotonation of a carboxylic acid is a reversible reaction. Consequently carboxylic acids can be separated readily from other classes of compounds which lack acidic functional groups, by virtue of their acid/base properties.

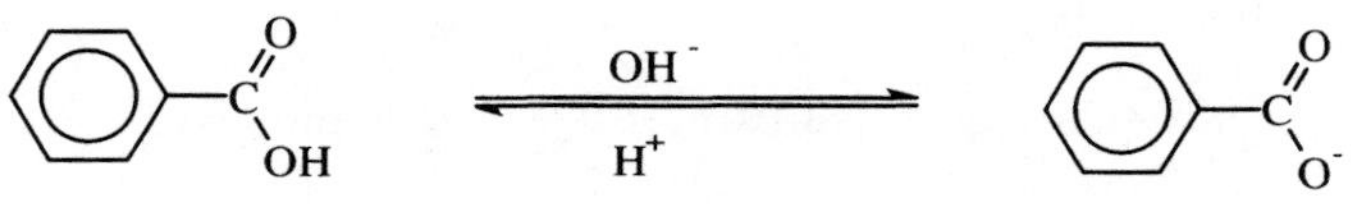

benzoic acid
(insoluble in water)

benzoate ion
(water soluble)

Carboxylic acid derivatives

There are several important classes of organic compounds which are derived from carboxylic acids. These include **esters, amides, acid anhydrides** and **acid chlorides**. All of the carboxylic acid derivatives can be synthesised from carboxylic acids, directly or indirectly, and a common reaction of all of the carboxylic acid derivatives is that they react with water (a hydrolysis reaction) to give the carboxylic acid as one of the products. Hydrolysis usually takes place with an acid or a base as a catalyst.

$$R-\overset{\overset{O}{\|}}{C}-Cl \qquad R = H, \text{ alkyl, aryl}$$

acid chloride

$$R-\overset{\overset{O}{\|}}{C}-O-\overset{\overset{O}{\|}}{C}-R' \qquad R, R' = H, \text{ alkyl, aryl}$$

acid anhydride

$$R-\overset{\overset{O}{\|}}{C}-O-R' \qquad R = H, \text{ alkyl, aryl} \\ R' = \text{alkyl, aryl}$$

ester

$$R-\overset{\overset{O}{\|}}{C}-\underset{\underset{R''}{|}}{N}-R' \qquad R, R', R'' = H, \text{ alkyl, aryl}$$

amide

Nitriles (compounds containing the —C≡N functional group) are usually considered along with esters, amides, and so on, as carboxylic acid derivatives because of the similarity of much of their chemistry and the fact that they also react with water to give carboxylic acids.

Esters, acid chlorides, amides, and anhydrides, vary considerably in their reactivity. Acid chlorides are much more reactive than other carboxylic acid derivatives and these, in turn, are more reactive than anhydrides. The general reactivity of the carboxylic acid derivatives follows the sequence:

acid chlorides > anhydrides > esters > amides

The general order of reactivity reflects some of the ways that the carboxylic acid derivatives can be made. Anhydrides, esters, and amides, can all be synthesised from acid chlorides. Esters and amides can be synthesised from acid chlorides and anhydrides. There are synthetic routes to amides from acid chlorides, anhydrides, and esters.

Acid chlorides are most easily prepared by the reaction of carboxylic acids with thionyl chloride ($SOCl_2$) or phosphorus trichloride (PCl_3). Acid chlorides are named by taking the name of the carboxylic acid and replacing the *-ic* ending of the acid name with *-yl* and adding the word *chloride*. The acid chloride derived from benzoic acid is called *benzoyl chloride*; the acid chloride derived from acetic acid is called *acetyl chloride*.

$$CH_3-\overset{\overset{O}{\|}}{C}-OH \xrightarrow{SOCl_2} CH_3-\overset{\overset{O}{\|}}{C}-Cl + HCl + SO_2$$

acetic acid → acetyl chloride

$$C_6H_5-\overset{\overset{O}{\|}}{C}-OH \xrightarrow{PCl_3} C_6H_5-\overset{\overset{O}{\|}}{C}-Cl$$

benzoic acid → benzoyl chloride

Acid chlorides are hydrolysed rapidly by water, even at room temperature, to give a carboxylic acid and HCl.

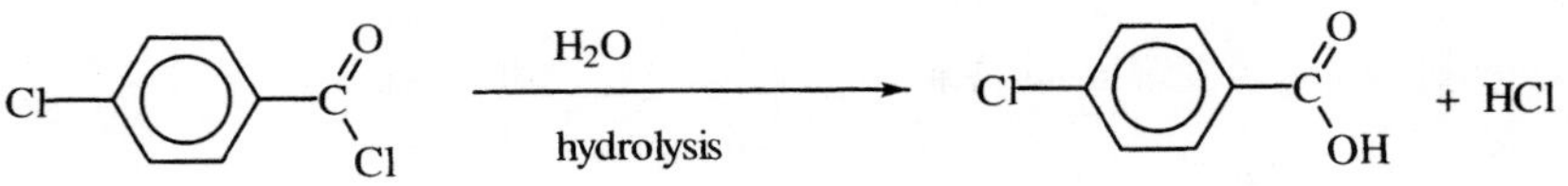

$$\text{p-chlorobenzoyl chloride} \xrightarrow[\text{hydrolysis}]{H_2O} \text{p-chlorobenzoic acid} + HCl$$

Some cyclic **acid anhydrides** can be formed directly by the high temperature dehydration of dicarboxylic acids, but, in general, acid anhydrides are formed by the reaction of acid chlorides with carboxylates.

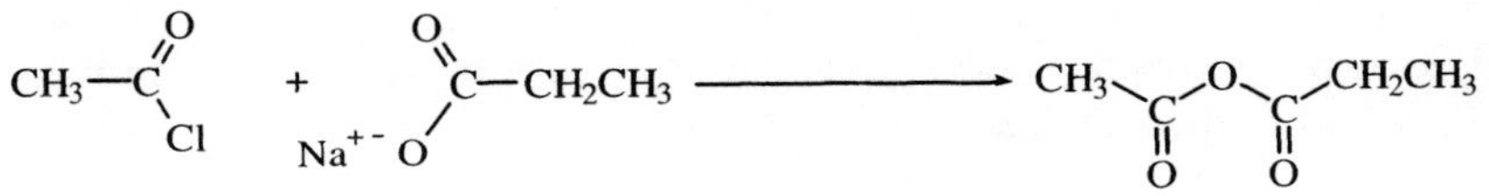

$$\text{acetyl chloride} + \text{sodium propionate} \longrightarrow \text{acetic propionic anhydride}$$

Symmetrical anhydrides are named simply by naming the carboxylic acid and adding the word *anhydride*. For unsymmetrical anhydrides, the carboxylic acids are named separately (in alphabetical order) and the word *anhydride* is added. An anhydride molecule can be considered to be constructed of two carboxylic acid molecules with a water molecule removed (hence the name). An anhydride is hydrolysed by water to give the carboxylic acids from which it was derived. Anhydrides are less reactive than acid chlorides and the hydrolysis reaction is slower. Nevertheless, anhydrides are readily hydrolysed.

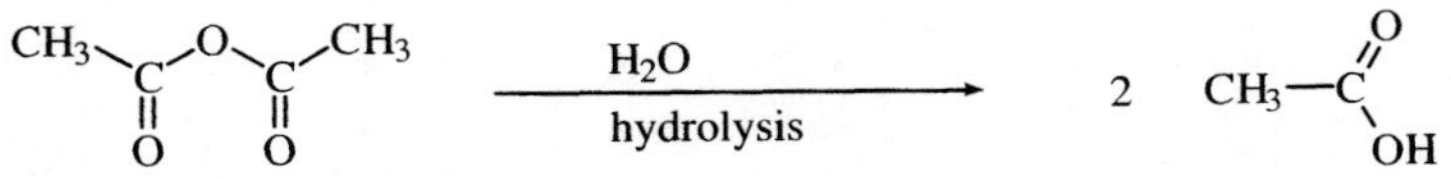

$$\text{acetic anhydride} \xrightarrow[\text{hydrolysis}]{H_2O} 2\ CH_3-COOH$$

Acid chlorides or anhydrides react with alcohols directly to form **esters**. The name of an ester consists of two words, which identify first the alcohol and second the carboxylic acid from which it was derived.

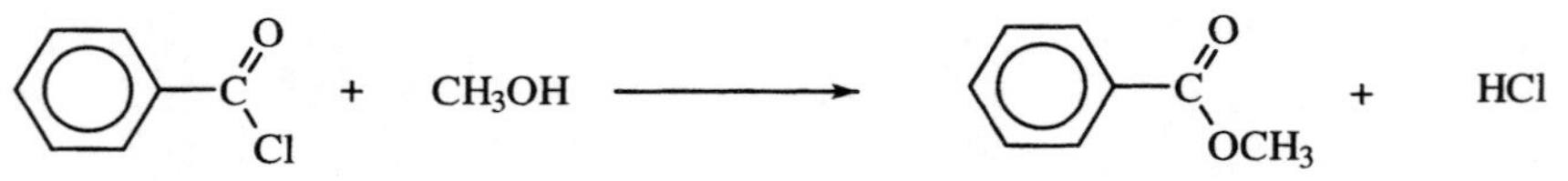

benzoyl chloride + CH₃OH → methyl benzoate + HCl

benzoyl chloride   methanol   methyl benzoate

acetic anhydride + CH₃CH₂OH → ethyl acetate + acetic acid

acetic anhydride   ethanol   ethyl acetate   acetic acid

Esters are hydrolysed by heating with dilute acid (H^+ / H_2O) or dilute base (OH^- / H_2O). On hydrolysis, esters give rise to a carboxylic acid (or carboxylate ion in basic solution) and an alcohol.

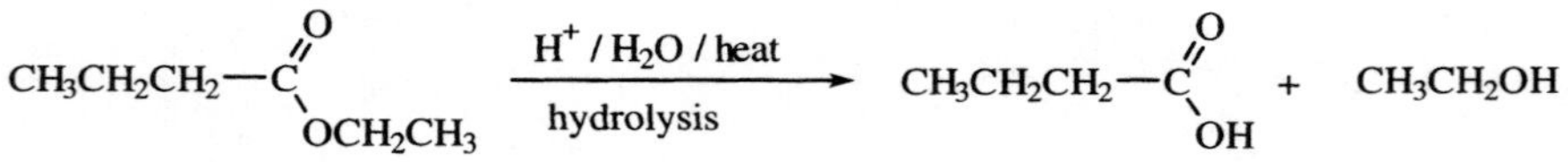

$$CH_3CH_2CH_2-C(=O)OCH_2CH_3 \xrightarrow[\text{hydrolysis}]{H^+ / H_2O / \text{heat}} CH_3CH_2CH_2-C(=O)OH + CH_3CH_2OH$$

ethyl butyrate   butyric acid   ethanol

Esters can be formed by the direct reaction of a carboxylic acid with an alcohol with the loss of H_2O. The reaction is typically catalysed by H^+ and because the water formed in the reaction can react in a hydrolysis reaction with the ester product, the acid and alcohol starting materials are in equilibrium with the ester and water products.

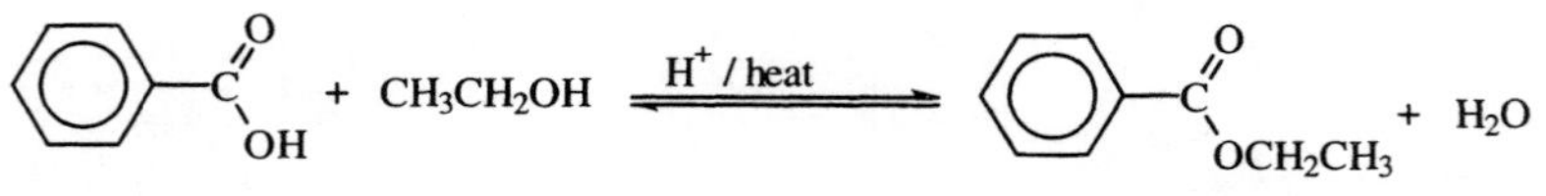

benzoic acid + CH₃CH₂OH ⇌ ethyl benzoate + H₂O (H⁺ / heat)

benzoic acid   ethanol   ethyl benzoate

The direct reaction between carboxylic acids and alcohols can be used to make esters in good yield providing the equilibrium can be driven to the right hand side of the equation by using a large excess of the acid or alcohol starting material and by removing either the water or the ester as it is formed.

Acid chlorides or anhydrides react with ammonia or with primary or secondary amines to form amides. Amides which have an acyl group and two hydrogen atoms attached to the nitrogen atom are called *primary* (1°) amides. Amides which have an acyl group, one alkyl group and one hydrogen atom attached to the nitrogen atom are called *secondary* (2°) amides, and amides with an acyl group and two alkyl groups attached to the nitrogen atom are called *tertiary* (3°) amides. Primary amides are named by identifying the carboxylic acid from which the amide is derived and replacing the -*ic* or -*oic* ending of the acid name with the ending -*amide*. The 1° amide derived from benzoic acid is called *benzamide*. The 1° amide derived from acetic acid is called *acetamide*. The stem of 2° and 3° amides is obtained in an analagous way and the substituents on the nitrogen indicated by a prefix to the stem and *N*- to indicate the point of attachment.

$$CH_3CH_2-\underset{Cl}{\overset{O}{\overset{\|}{C}}} \;+\; NH_3 \longrightarrow CH_3CH_2-\underset{NH_2}{\overset{O}{\overset{\|}{C}}} \;+\; HCl$$

propionyl chloride   ammonia   propionamide
                                                      (a primary amide)

$$C_6H_5-\underset{Cl}{\overset{O}{\overset{\|}{C}}} \;+\; CH_3NH_2 \longrightarrow C_6H_5-\underset{\underset{H}{|}}{\overset{O}{\overset{\|}{C}}}-N-CH_3 \;+\; HCl$$

benzoyl chloride   methylamine   *N*-methylbenzamide
                         (a primary amine)   (a secondary amide)

$$CH_3-\overset{O}{\overset{\|}{C}}-O-\overset{O}{\overset{\|}{C}}-CH_3 \;+\; H-\underset{CH_3}{\overset{CH_3}{\overset{|}{N}}} \longrightarrow CH_3-\underset{\underset{CH_3}{|}}{\overset{O}{\overset{\|}{C}}}-N-CH_3 \;+\; CH_3-\underset{OH}{\overset{O}{\overset{\|}{C}}}$$

acetic anhydride   dimethylamine   N,N-dimethylacetamide   acetic
                     (a secondary amine)   (a tertiary amide)   acid

Amides are hydrolysed by heating with aqueous acid or dilute base. Amides are the most difficult of the carboxylic acid derivatives to hydrolyse and usually require prolonged heating with more concentrated acid or base than is necessary for the hydrolysis of esters, anhydrides, or acid chlorides. On hydrolysis, amides give rise to a carboxylic acid (or carboxylate ion in basic solution) and an amine (or protonated amine in acid solution):

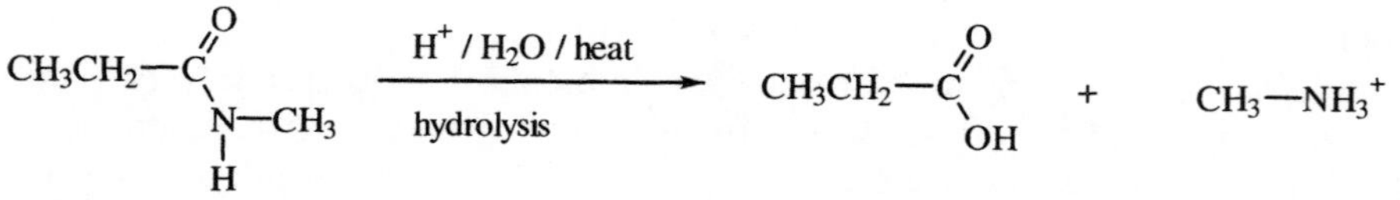

CH₃CH₂—C(=O)N(H)(CH₃) →[H⁺ / H₂O / heat, hydrolysis] CH₃CH₂—C(=O)OH + CH₃—NH₃⁺

N-methylpropionamide   propionic acid   methylammonium ion

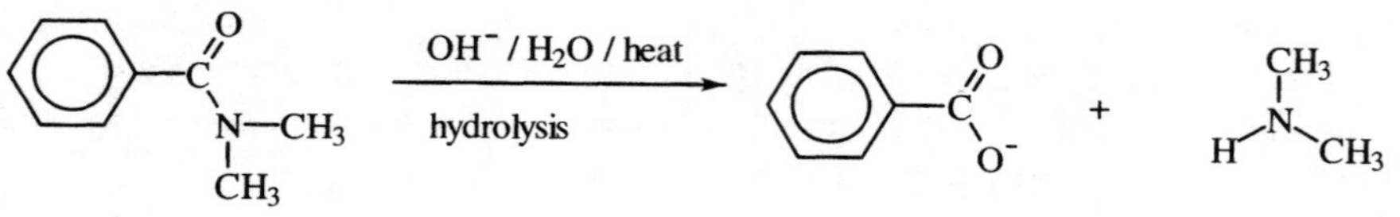

C₆H₅—C(=O)N(CH₃)(CH₃) →[OH⁻ / H₂O / heat, hydrolysis] C₆H₅—C(=O)O⁻ + HN(CH₃)(CH₃)

N,N-dimethlbenzamide   benzoate ion   dimathylamine

Proteins are among the most important natural amides and are polymeric materials formed from amino acids. Amino acids have both an amine functional group and a carboxylic acid functional group in the same molecule and proteins are formed by linking the amine group from one molecule with the carboxylic acid group of another to form an amide bond. Proteins generally consist of hundreds of amino acid residues linked together to form a single molecule with a molecular weight of $\approx 10^5$.

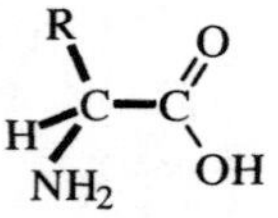

an amino acid

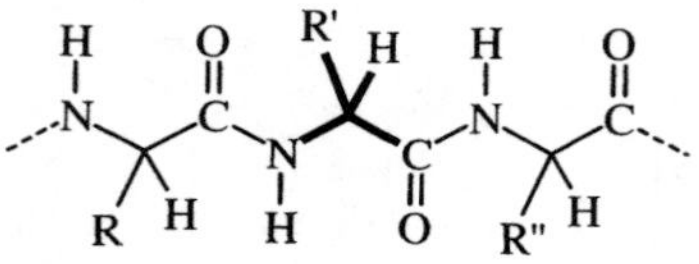

a protein—a polyamide
formed from amino acids

Hydrolysis of a protein is typically achieved by treating it with boiling 6M hydrochloric acid. This hydrolyses all of the amide bonds and releases all of the amino acids which make up the protein.

Formation of carboxylic acids

Carboxylic acids can be formed in three important ways: (i) hydrolysis of any of the carboxylic acid derivatives (including nitriles); (ii) oxidation of primary alcohols (Section 2.2) or aldehydes (Section 2.6); recall also that the alkyl side chains of alkylaromatic compounds can be oxidised to give aromatic carboxylic acids (Section 1.3); (iii) reaction of Grignard reagents with carbon dioxide.

The C=O bond of carbon dioxide is similar to that of the carbonyl group of the aldehydes and ketones, and Grignard reagents add to the carbon atom to form a carboxylate anion. Reaction of the carboxylate with H⁺ produces the free carboxylic acid. This is a particularly useful method for extending the carbon chain of a compound by one carbon.

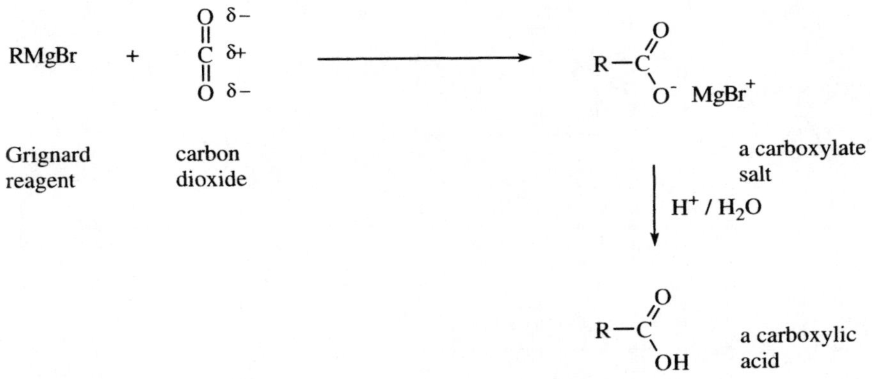

$$RMgBr + CO_2 \longrightarrow R-C(=O)-O^- \; MgBr^+$$

Grignard reagent   carbon dioxide   a carboxylate salt

$$\downarrow H^+ / H_2O$$

$$R-C(=O)-OH \quad \text{a carboxylic acid}$$

SAMPLE EXERCISE 2.15

Devise a synthetic scheme which would convert 1-bromobutane to pentanoic acid.

Solution: Firstly note that the starting material has four carbon atoms and the product has five so one carbon atom must be added during the course of the synthesis. Since the product is a carboxylic acid, the synthesis is best performed by reaction of a Grignard reagent with CO_2. The alkyl halide is reacted with magnesium in dry ether solvent to form butylmagnesium bromide (a Grignard reagent) and this is treated with CO_2 to form a pentanoate salt. When the reaction mixture is acidified with dilute acid, pentanoic acid is the product.

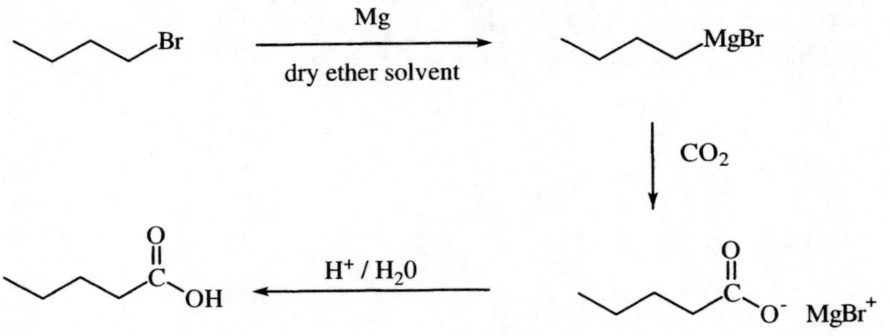

$$\text{CH}_3\text{CH}_2\text{CH}_2\text{CH}_2\text{Br} \xrightarrow[\text{dry ether solvent}]{Mg} \text{CH}_3\text{CH}_2\text{CH}_2\text{CH}_2\text{MgBr}$$

$$\downarrow CO_2$$

$$\text{CH}_3\text{CH}_2\text{CH}_2\text{CH}_2\text{-C(=O)-OH} \xleftarrow{H^+ / H_2O} \text{CH}_3\text{CH}_2\text{CH}_2\text{CH}_2\text{-C(=O)-O}^- \; MgBr^+$$

Reduction of carboxylic acids and esters to primary alcohols

Both carboxylic acids and esters can be reduced to primary alcohols using lithium aluminium hydride (in a dry reaction solvent) as a reducing agent. The reaction is similar to the reduction of aldehydes (Section 2.6) and the immediate product is the alkoxide of the primary alcohol which gives the primary alcohol on reaction with dilute acid.

Carboxylic acids are generally difficult to reduce and often require prolonged heating to complete the reduction; however esters are reduced easily and, in practice, it is often more convenient to convert the carboxylic acid to the ester before performing the reduction.

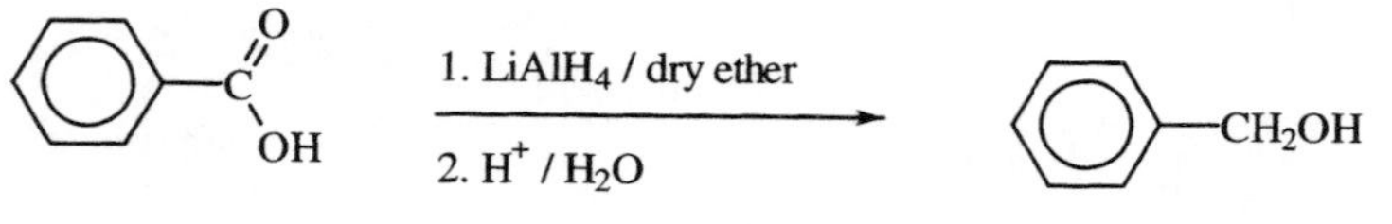

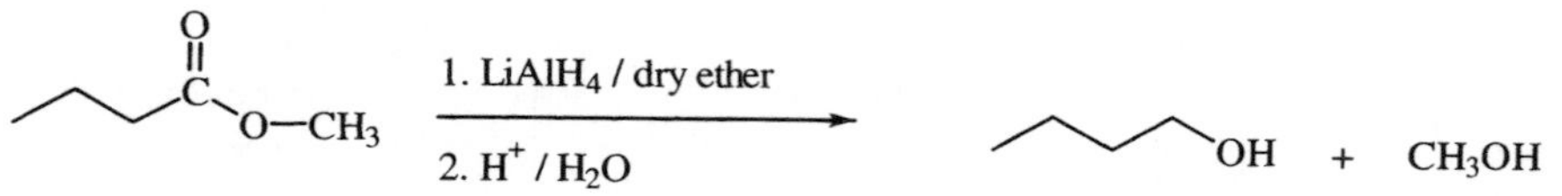

Carboxylic Acids Around Us

Prostaglandins form a family of hormone-like compounds, based on *eicosatrienoic acid*, which regulate many bodily functions, such as the reproductive cycle, growth, and blood clotting. In different circumstances prostaglandins can cause or suppress inflammation. Prostaglandin synthesis is stimulated by tissue damage and is responsible for some of the pain associated with cuts and bruises. Aspirin blocks this synthesis and so results in pain relief. The effect of prostaglandins was first noted in 1930 by Raphael Kurzrok and Charles Lieb who observed that human semen stimulated contraction of isolated uterine muscle. The prostaglandin derivatives PGE_2 and $PGF_{2\alpha}$ can be used to induce abortion during the second trimester of pregnancy.

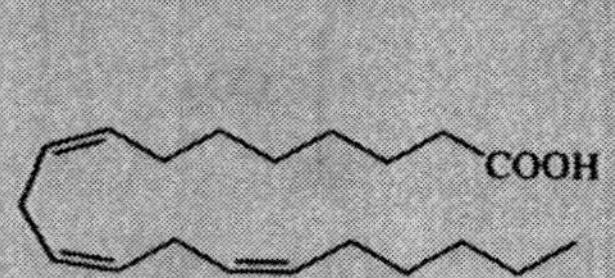

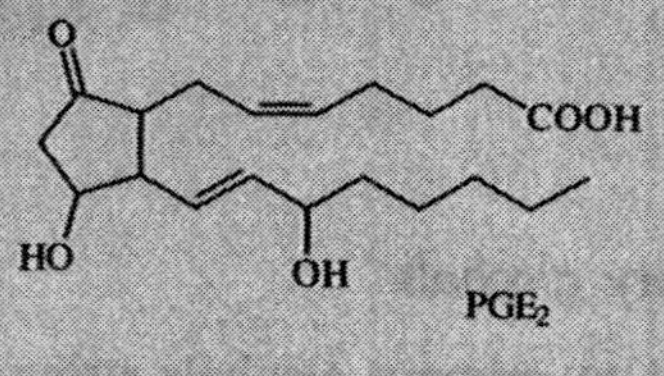

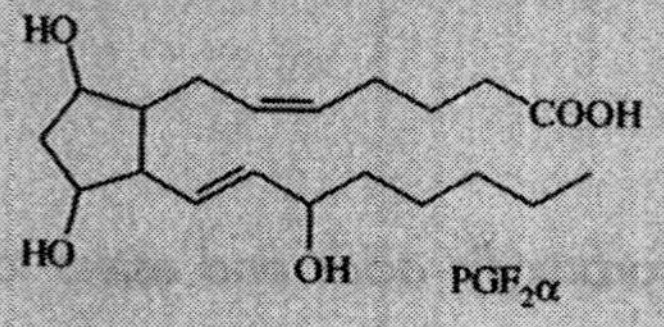

Chrysanthemic acid: Commonly known as *pyrethrin*, the structure of this carboxylic acid was proved in 1920 by Leopold Ruzicka, while working in Switzerland. He won a Nobel Prize for his work on the class of natural products called *terpenes*, of which chrysanthemic acid is a member. It is isolated from an East African chrysanthemum and is an insecticide.

chrysanthemic acid

Salicin: This is present in the leaves and bark of willow trees and has been used for hundreds of years in a variety of herbal remedies. It is converted into salicylic acid *in vivo* which acts as an agent to reduce inflammation and lower the temperature of patients suffering from fever. However, large doses of salicylic acid taste unpleasant and also cause gastric irritation. This was largely overcome, in 1899, by the introduction of acetylsalicylic acid or *Aspirin*. This is an ester which passes through the stomach unchanged before being hydrolysed by the basic medium of the intestine to form the active compound.

salicin   salicilic acid   acetylsalicylic acid

SUMMARY

Most organic compounds contain additional elements to carbon and hydrogen, and are classified on the basis of the functional groups these elements form.

Organic compounds with a halogen attached to an aliphatic chain are called *alkyl halides* to distinguish them from the *aryl halides* in which the halogen is attached to an aromatic ring. The relatively high electronegativity of the halogen atoms results in a polar C—X bond with a partial positive charge residing on the carbon atom and a partial negative charge on the halogen. The chemistry of the alkyl halides is dominated by the attack of nucleophiles on this carbon atom; however, alkyl fluorides rarely undergo nucleophilic substitution due to the C—F bond being relatively strong. Under strongly basic conditions HX can be eliminated from an alkyl halide to give an alkene. Alkyl halides and aryl halides react with magnesium to produce highly reactive and useful compounds known as *Grignard reagents*.

Alcohols are compounds with an —OH group attached to an sp^3 hybridised carbon atom. The O—H bond of alcohols is strongly polarised and the atoms are capable of hydrogen bonding. As a consequence of these intermolecular interactions, alcohols have relatively high boiling points and the smaller alcohols are soluble in water. The proton of the O—H group can be removed by a strong base to give an alkoxide ion. Alcohols are readily oxidised by oxidising agents such as MnO_4^- or $Cr_2O_7^{2-}$. Primary alcohols are oxidised via an aldehyde to a carboxylic acid, secondary alcohols are oxidised to ketones. Alcohols can be converted to alkyl halides by nucleophilic substitution and to alkenes by dehydration. Compounds in which the —OH group is attached to an aromatic ring are called *phenols*. Like alcohols, phenols can be deprotonated, in this case, to give a phenoxide ion.

Ethers have an oxygen atom attached to two alkyl or aryl groups. Ethers are excellent solvents because of their stability and lack of reactivity, but they can be converted to alkyl iodides by reaction with concentrated hydroiodic acid.

Amines are weak bases as a consequence of the lone pair of electrons residing on the sp^3 hybridised nitrogen atom. N—H bonds in amines are highly polarised and, like alcohols, can form hydrogen bonds and are relatively soluble in water. Amines are readily protonated to give ammonium cations. Amines can be alkylated with alkyl halides. Aromatic amines can be prepared by reducing aromatic nitro compounds. Aromatic amines react with HNO_2 at low temperature to produce *aromatic diazonium cations* which are useful synthetic intermediates for producing nitriles, aryl halides, phenols, and azo compounds.

Compounds that contain a carbonyl (C=O) group are classified as *aldehydes* if the carbon atom has at least one hydrogen atom attached and *ketones* if the carbon has two alkyl or aryl groups attached. Nucleophiles attack the carbon atom of the C=O group in a nucleophilic addition reaction to give alcohols; amine nucleophiles give imines, water gives geminal diols, and alcohols give hemiacetals. Simple sugars, such as glucose, form cyclic hemiacetals. Aldehydes are readily oxidised to carboxylic acids and aldehydes and ketones can be catalytically reduced to alcohols.

Carboxylic acids contain the carboxyl (—COOH) group, the most acidic organic functional group. When deprotonated, this group (—COO⁻) is called a *carboxylate* ion. Carboxyl groups are highly polarised and participate in strong hydrogen bonds. There are a number of important carboxylic acid derivatives, including *esters*, *amides*, *acid anhydrides*, and *acid chlorides*. The chlorides and anhydrides are particularly reactive and are useful as reagents. Esters and amides can be hydrolysed by heating with dilute acid or dilute base. Carboxylic acids and esters can be reduced to primary alcohols using a powerful reducing agent, such as lithium aluminium hydride.

KEY TERMS

acetal
acid anhydride
acid chloride
alcohol
aldehyde—alkanal
alkoxide ion
alkyl halide
amide
amine
ammonium cation
aromatic diazonium salt
aryl halide
carbonyl
carboxylic acid

diol
elimination reaction
ester
ether
Grignard reagent
hemiacetal
hydrogen bond
imine
ketone—alkanone
nitrile
nucleophile
nucleophilic substitution
phenol
phenoxide

EXERCISES

Alkyl halides

1. Classify each of the following compounds as a primary, secondary, or tertiary alkyl halide or an aryl halide.

    (a) CH_3CH_2-F

    (b) 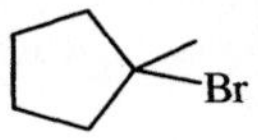 cyclopentyl-Br

    (c) 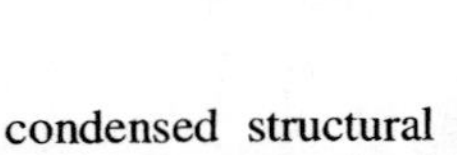 Cl—⟨C$_6$H$_4$⟩—CH_3

    (d) 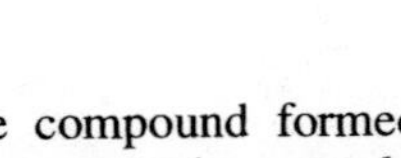 ⟨C$_6$H$_5$⟩—CH_2Br

    (e) 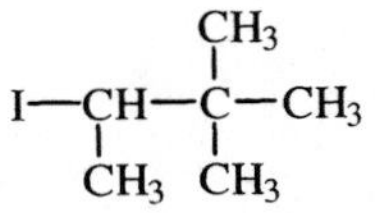 $I-CH(CH_3)-C(CH_3)_2-CH_3$

2. Give the condensed structural formula and name of the compound formed when 2-bromopropane is treated with each of the following reagents in a nucleophilic substitution reaction: (a) sodium hydroxide in dilute aqueous solution; (b) triethylamine; (c) potassium acetate (CH_3COOK); (d) sodium propynide ($CH_3C\equiv CNa$)

3. Give the condensed structural formula and name of the compound formed when iodocyclopentane is treated with each of the following reagents in a nucleophilic substitution reaction: (a) sodium hydroxide in dilute aqueous solution; (b) sodium methoxide (c) sodium cyanide (NaCN); (d) potassium butyrate ($CH_3CH_2CH_2COOK$).

4. Give the product that would be formed when the following compound is treated with sodium methoxide in a nucleophilic substitution reaction:

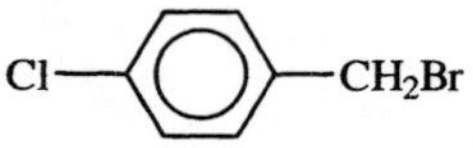

Cl—⟨⟩—CH₂Br

5. Give the products that would be formed when each of the following compounds undergoes an elimination reaction after treatment with a hot, concentrated solution of potassium hydroxide in alcohol solvent. If more than one product is formed indicate which is the major product.

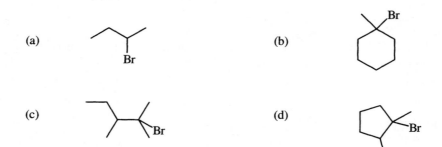

6. (a) What are the reagents and conditions required to form pentylmagnesium bromide?
(b) What is the organic product formed when this Grignard reagent reacts with dilute hydrochloric acid?

7. Devise a reaction scheme to convert 2-bromopentane to pentane.

Alcohols

8. Classify each of the following compounds as a primary, secondary, or tertiary alcohol, or a phenol.

(a) $CH_3(CH_2)_4CH_2OH$

(b) 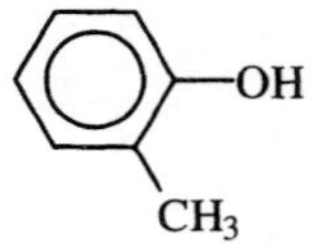 [o-cresol structure with OH and CH₃]

(c) 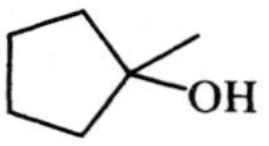 [cyclopentane with OH]

(d) 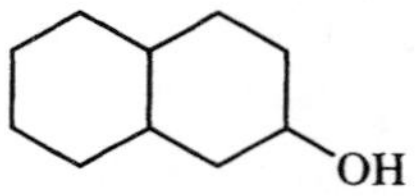 [decalin with OH]

(e) 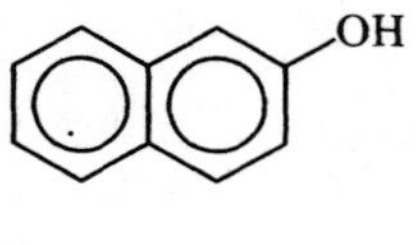 [naphthalene with OH]

9. Give the condensed structural formula and name of the compound formed when 2-propanol is treated with sodium metal and then with bromomethane.
10. Give the condensed structural formula of the compound formed when cyclopentanol is treated with sodium amide and then with 2-bromopropane.
11. Give the condensed structural formula and name of the compound formed when 2-pentanol is treated with each of the following reagents: (a) phosphorus trichloride; (b) phosphorus tribromide; (c) potassium metal; (d) thionyl chloride.
12. Give two different methods for performing the following transformation:

13. Give the products that would be formed when each of the following compounds undergoes a dehydration reaction after treatment with concentrated sulfuric acid. If more than one product is formed indicate which you would expect to be the major product.

(a) $\begin{array}{c}CH_3 \\ \diagdown \\ CH-CH_2OH \\ \diagup \\ CH_3\end{array}$

(b) cyclopentanol

(c) (CH₃)₃C—CH(CH₃)—OH (neopentyl-type with OH on secondary carbon)

(d) (CH₃)₃C—CH(OH)—CH₃

14. Give the products that would be formed when each of the following compounds is oxidised with potassium dichromate in acidic solution. If no reaction occurs write 'no reaction':

(a) 2-pentanol (CH₃CH₂CH₂CH(OH)CH₃)

(b) CH₃(CH₂)₂CH₂OH

(c) 2-methylcyclohexanol (OH on carbon adjacent to methyl)

(d) 1-methylcyclohexanol (tertiary)

15. Give the products that would be formed when each of the following compounds is oxidised with potassium dichromate in acidic solution. If no reaction occurs, write 'no reaction': (a) 2-methyl-2-butanol; (b) 2-methyl-1-butanol; (c) 2,2-dimethyl-3-butanol; (d) 3,3-diethylcyclohexanol.

16. Devise a synthetic scheme for converting cyclopentene to cyclopentanone.
17. Devise a synthetic scheme for converting cyclopentanol to cyclopentane.

Ethers and Phenols

18. Describe a method you could use to separate *p*-cresol (*p*-hydroxytoluene) from chlorobenzene using the acid/base properties of the compounds.
19. What are the products that would result when each of the following compounds is treated with concentrated hydriodic acid?

    (a)   butyl methyl ether   (b)  tetrahydrofuran

    (c) 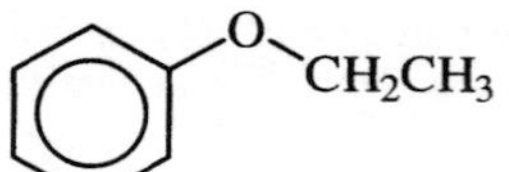 phenyl ethyl ether (C$_6$H$_5$-O-CH$_2$CH$_3$)

20. What is the product that would result when *p*-chlorophenol is treated with sodium hydroxide and then with 2-bromobutane?

Amines

21. Describe a method you could use to separate aniline from chlorobenzene using the acid/base properties of the compounds.
22. Give the condensed structural formulas of the products obtained when each of the following compounds is treated with an excess of 1-iodopropane.

    (a)   CH$_3$—NH$_2$ (b) 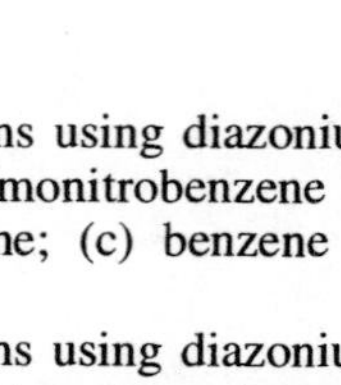 *N*,*N*-dimethylaniline (C$_6$H$_5$N(CH$_3$)$_2$)

    (c) (CH$_3$)$_2$NH

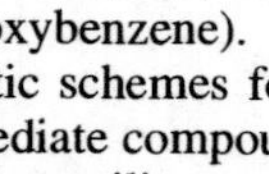

23. Devise synthetic schemes for performing the following conversions using diazonium salts as intermediate compounds in the syntheses: (a) *p*-bromonitrobenzene to *p-bromobenzonitrile;* (b) 2,6-dichloroaniline to *m*-dichlorobenzene; (c) benzene to anisole (methoxybenzene).
24. Devise synthetic schemes for performing the following conversions using diazonium salts as intermediate compounds in the syntheses: (a) *p*-nitrotoluene to *p*-bromobenzoic acid; (b) *o*-chloroaniline to *o*-chloroiodobenzene; (c) benzene to phenol.

Aldehydes and ketones

25. Devise synthetic schemes for performing the following conversions using Grignard reagents as intermediate compounds in the syntheses: (a) 2-bromobutane to 2-methyl-1-butanol; (b) 1-chloropropane to 2-methyl-2-pentanol; (c) 2-iodopropane to 3-methyl-2-butanol.

26. Devise a synthetic scheme for synthesising each of the following alcohols:

    (a) 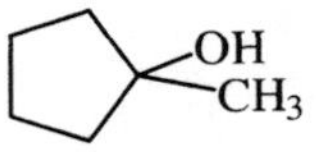 1-methylcyclopentanol (cyclopentane with −OH and −CH₃)

    (b) 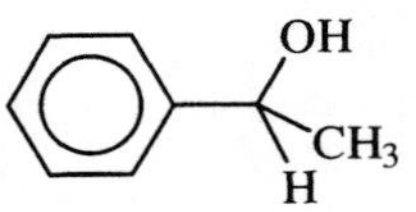 1-phenylethanol (benzene ring with −C(OH)(CH₃)H)

    (c) 2-phenyl-2-methylpropan-2-ol type structure (benzene−CH₂−C(OH)(CH₃)−CH₂CH₃)

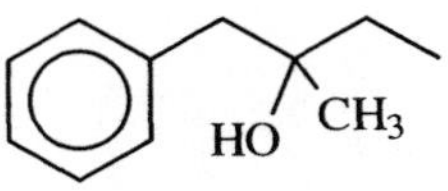

27. Give the condensed structural formulas of the products that would be obtained when each of the following compounds is treated first with lithium aluminium hydride in dry ether solvent then with dilute HCl: (a) 2-butanone; (b) p-chlorobenzaldehyde; (c) 3,3-dimethylcylopentanone.

28. Give the condensed structural formulas of the products that would be obtained when each of the following compounds is treated first with sodium borohydride in methanol solvent then with dilute HCl.

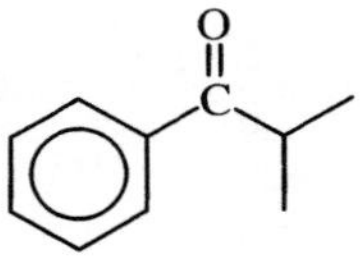

    (a) 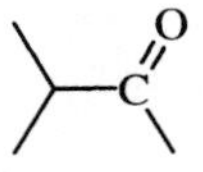 (CH₃)₂CH−C(=O)−CH₃ type structure

    (b) phenyl−C(=O)−CH(CH₃)₂

    (c) 1-indanone (bicyclic structure with C=O)

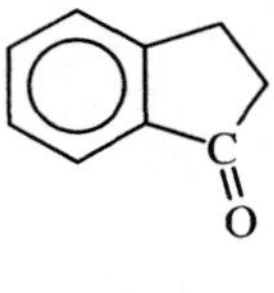

29. Give the condensed structural formulas of the products that would be obtained in each of the following reactions and give the class name of the product which is formed: (a) acetone is treated with semicarbazide; (b) acetophenone is treated with 2,4-dinitrophenylhydrazine; (c) aniline is treated with 4-methoxybenzaldehyde.

30. Give the condensed structural formulas of the products that would be obtained in each of the following reactions and give the class name of the product which is formed: (a) 3-pentanone is treated with hydroxylamine; (b) cyclopentanone is treated with hydrazine; (c) 3,3-dimethylbutyraldehyde is treated with semicarbazide.

31. Give the condensed structural formula of the product that would be obtained when cyclopentanone is treated with ethylene glycol (1,2-ethanediol) in the presence of HCl.

32. Give the condensed structural formula of the product that would be obtained when 1,3-propandiol reacts with acetone in the presence of HCl.

33. Give the condensed structural formula of the β-hydroxyaldehyde that results when butyraldehyde undergoes an aldol condensation in the presence of dilute sodium hydroxide.
34. Give the condensed structural formula of the α,β-unsaturated ketones that result from the dehydration of the intermediate β-hydroxyketones formed by an aldol condensation of the following compounds. Indicate whether more than one stereoisomer is possible for the product(s).

(a) 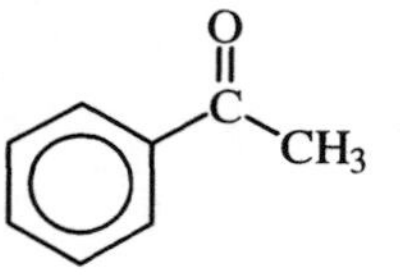 C$_6$H$_5$-CO-CH$_3$ (acetophenone)

(b) CH$_3$CH$_2$-CO-CH$_2$CH$_3$ (3-pentanone)

(c) CH$_3$-CO-CH$_2$CH$_3$ (2-butanone)

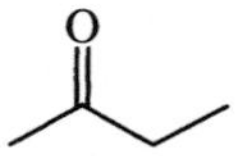

35. Give the condensed structural formulas of the two β-hydroxyaldehyde products which are possible from aldol condensations when acetaldehyde is mixed with propionaldehyde in the presence of a base.
36. Tetracyclone is formed by the reaction of benzil with 1,3-diphenylacteone in the presence of a base. Rationalise how tetracyclone could form.

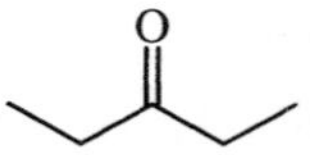

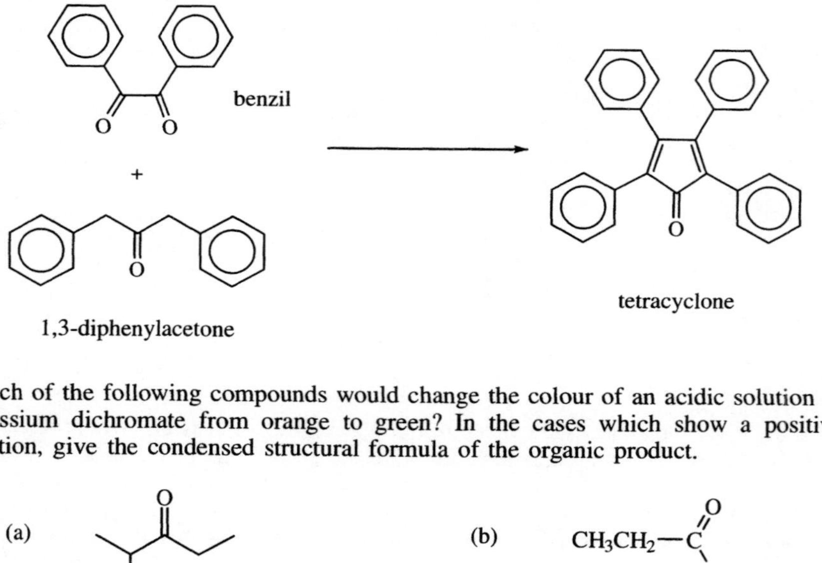

benzil + 1,3-diphenylacetone → tetracyclone

37. Which of the following compounds would change the colour of an acidic solution of potassium dichromate from orange to green? In the cases which show a positive reaction, give the condensed structural formula of the organic product.

(a) 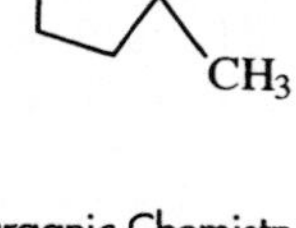 (CH$_3$)$_2$CH-CO-CH$_2$CH$_3$ type structure

(b) CH$_3$CH$_2$-C(=O)-H

(c) cyclopentane with OH and CH$_3$ substituents on same carbon

38. Which of the following compounds would give a precipitate of metallic silver (a silver mirror) from a basic solution of [Ag(NH$_3$)$_2$]$^+$? In the cases which show a positive reaction, give the condensed structural formula of the organic product.

(a) 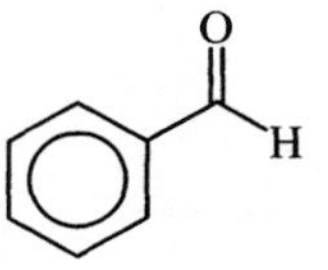 benzaldehyde (C$_6$H$_5$CHO)

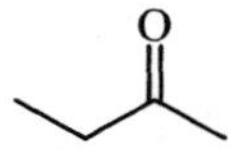

(b) methyl propyl ketone / butan-2-one structure

(c) butanal (CH$_3$CH$_2$CH$_2$CHO)

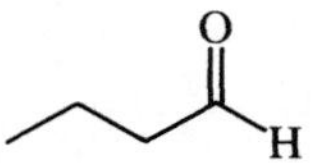

(d) 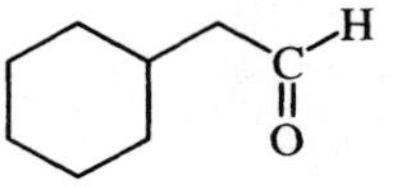 cyclohexylacetaldehyde

Carboxylic acids and derivatives

39. Devise reaction schemes which would allow you to: (a) convert benzoic acid and 1-propanol to 1-propyl benzoate; (b) propionic acid and methylamine to N-methylpropionamide; (c) 2-methylbutyric acid and ethanol to ethyl 2-methylbutyrate.

40. Give the names and condensed structural formulas of the organic products that would be obtained in the following reactions: (a) acetyl chloride is treated with 2-propanol; (b) benzoyl chloride is treated with dimethylamine; (c) acetic anhydride is treated with phenol; (d) butyric anhydride is treated with 1-butanol; (e) p-chlorobenzoic acid is treated firstly with thionyl chloride and then with cyclopentanol.

41. Give the names and condensed structural formulas of the organic products that would be obtained when each of the following compound is hydrolysed by heating with 6M HCl.

(a) 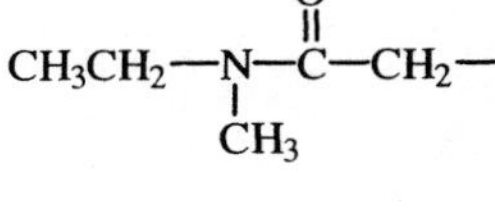 CH$_3$CH$_2$C(=O)OCH$_2$CH$_3$

(b) C$_6$H$_5$C(=O)O-C$_6$H$_5$

(c) cyclopentyl-O-C(=O)CH$_3$

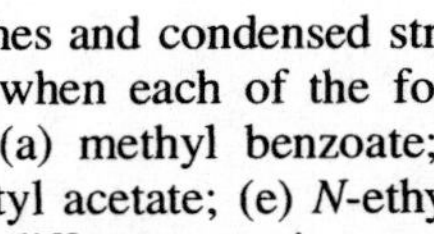

(d) 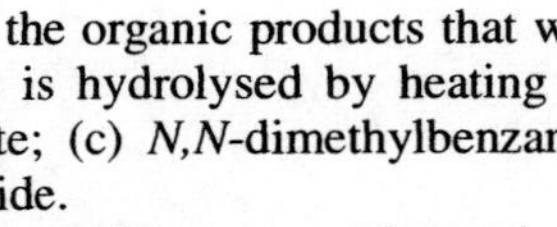 C$_6$H$_5$-N(H)-C(=O)-CH$_2$CH$_3$

(e) 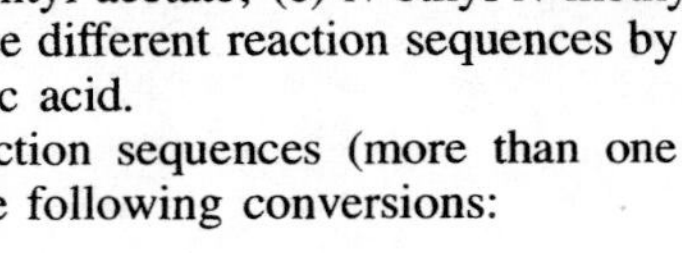 CH$_3$CH$_2$-N(CH$_3$)-C(=O)-CH$_2$-cyclohexyl

42. Give the names and condensed structural formulas of the organic products that would be obtained when each of the following compounds is hydrolysed by heating with 6M NaOH: (a) methyl benzoate; (b) phenyl butyrate; (c) N,N-dimethylbenzamide; (d) cyclopentyl acetate; (e) N-ethyl-N-methylbutyramide.

43. Devise three different reaction sequences by which you could convert 1-bromohexane to heptanoic acid.

44. Devise reaction sequences (more than one step is necessary) by which you could perform the following conversions:

Derivatives of Hydrocarbons **123**

(a) 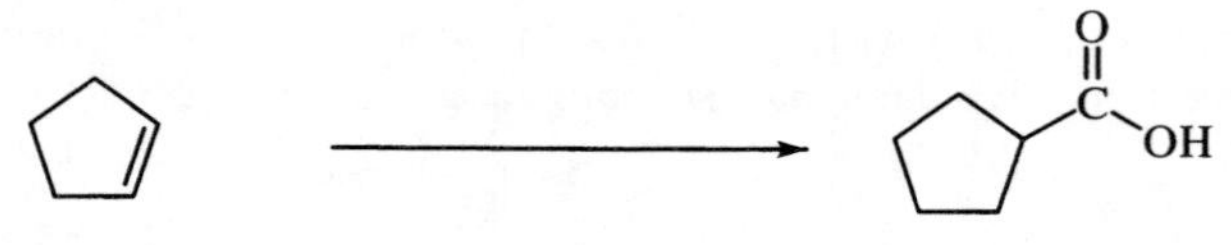 cyclopentene ⟶ cyclopentanecarboxylic acid

(b) 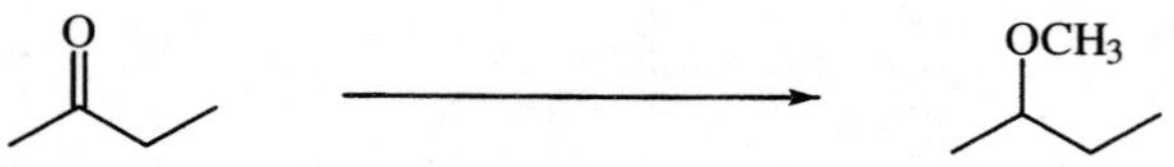 butan-2-one ⟶ 2-methoxybutane

(c) 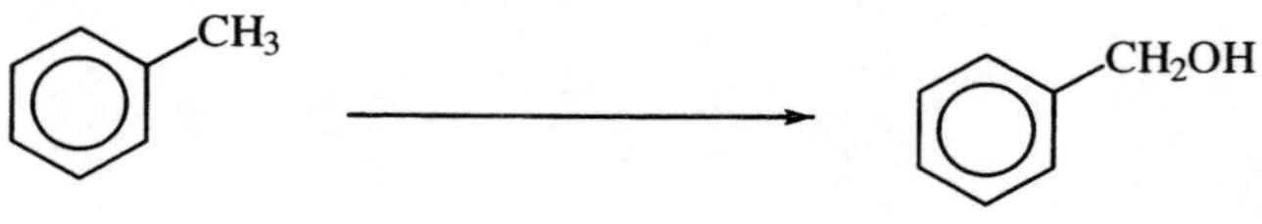 toluene ⟶ benzyl alcohol

45. Devise reaction sequences (more than one step is necessary) by which you could perform the following conversions:

(a) 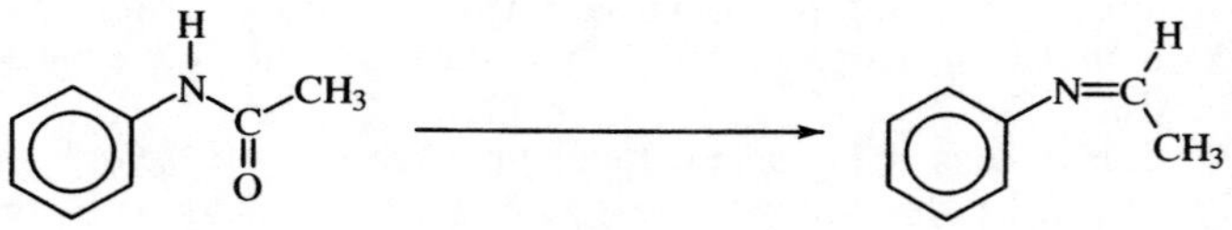 C₆H₅–NH–C(=O)–CH₃ ⟶ C₆H₅–N=CH–CH₃

(b) butan-2-ol ⟶ 2-methylbutan-2-ol

(c) benzonitrile ⟶ benzamide

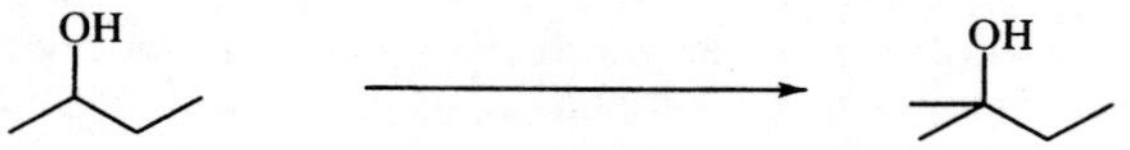

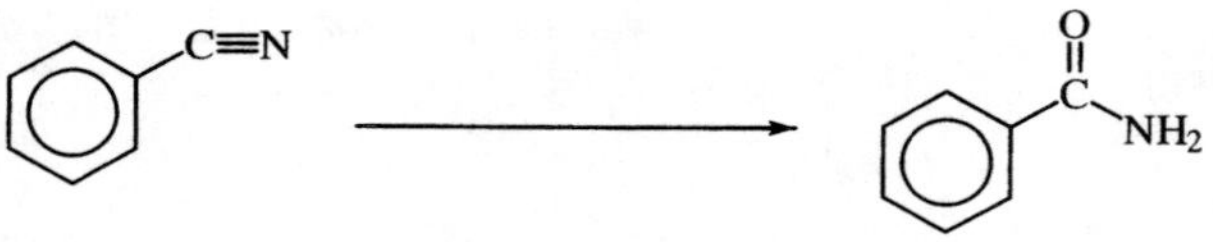

Separation and Identification of Organic Compounds 3

From Chapters 1 and 2 it is clear that given the possible variations in the carbon skeleton and presence of one or more of the common functional groups, there is effectively an infinite number of possible organic compounds. Each year many thousands of new or previously unknown organic compounds are prepared in the laboratory or extracted from natural sources. These must be identified and fully characterised and, in order to identify an unknown organic compound, or the components of a mixture of organic compounds, it is necessary firstly to separate and purify the compounds. In this chapter we describe firstly the most important **separation** and purification techniques and second, the major **identification** techniques.

3.1 SEPARATION AND PURIFICATION OF ORGANIC COMPOUNDS

Separation of a mixture of liquids or purification of a liquid is usually achieved by **distillation**, providing the liquids are sufficiently volatile. Distillation simply involves heating the mixture of liquids carefully until the temperature reaches the boiling point of one of the components (Figure 3.1). This component then distils from the reaction mixture and is collected. Each volatile component of the mixture can be removed from the reaction mixture at a temperature near its boiling point. If the difference in the boiling points of some of the components of the mixture is not sufficiently large to permit a clean separation, the products of the distillation can be re-distilled to improve their purity, or a **fractionating column** (Figure 3.1) can be used to enhance the performance of the distillation. A fractionating column consists of a series of surfaces on which the vaporised liquids can condense and re-evaporate after leaving the main distillation pot. The liquid is effectively distilled many times by condensation and re-vaporisation and this substantially improves the resolution with which liquids can be separated by distillation.

If the boiling points of the liquids to be purified are too high to be reached easily in a laboratory or if the compounds are unstable and begin to decompose when heated near their boiling points, distillation can be performed at reduced pressure. Reducing the pressure inside the distillation apparatus (by using a vacuum pump) reduces the boiling points of the components.

Separation of a mixture of solids or purification of solids is best achieved by **crystallisation** and **recrystallisation**. This process involves identifying a suitable solvent in which the material to be recrystallised or purified has high solubility at elevated temperature (typically near the boiling point of the solvent) and low solubility at low temperature. The solid (or mixture) is dissolved in a minimal volume of hot solvent, filtered to remove any insoluble components in the mixture and the solution is cooled slowly to allow the solute to crystallise. The formation and growth of crystals is a complex process, but involves the removal of molecules from solution (molecule by molecule) and stacking them into a regular condensed state where the molecules

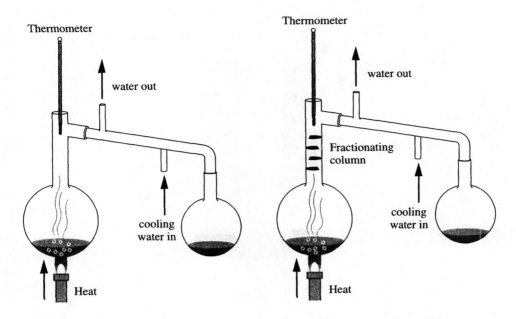

Figure 3.1 Standard distillation apparatus and distillation apparatus with fractionating column

are packed in a crystal lattice. Like molecules tend to pack more efficiently so the growth of crystals tends to bring together molecules of the same type. The effect of dissolving a solid then permitting it to crystallise is that the crystals which form contain less impurities (defects) than the material which was originally dissolved. Successive recrystallisation can produce extremely pure crystalline samples. Crystallisation is the purification method of choice where the solid is highly crystalline and forms crystals readily. There is considerable flexibility in selecting the appropriate solvent or mixture of solvents for the crystallisation of a particular compound and an experienced organic chemist quickly develops the skill which permits purification of organic compounds by crystallisation. Some materials, particularly those containing long hydrocarbon chains, do not form crystals readily and these must be purified by other means.

Some solids can be purified by **sublimation**. This is the analogue of distillation of liquids where the solid is heated, typically under a highly reduced pressure (10^{-3} to 10^{-6} mm Hg) to produce a gas (Figure 3.2). Some solids evaporate directly into the gas phase without melting into liquids and the gas can then be condensed back to a solid on a cold surface. The method is only applicable to solids where the solid-to-gas phase transition can be reached at temperatures and pressures accessible in the laboratory. Because this separation/purification method can be used only for a limited range of solids it is a particularly useful method for isolating compounds which can be sublimed from those which cannot.

Amines and carboxylic acids can be separated from compounds which do not contain basic or acidic functional groups by **acid/base extraction**. Amines are organic

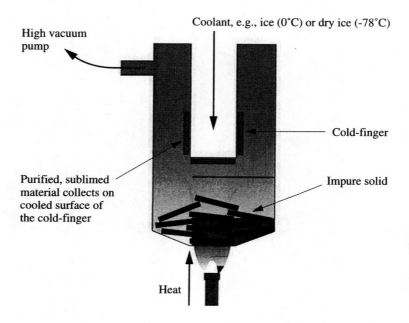

Figure 3.2   Sublimation apparatus

bases and can be protonated by acids to form cationic, substituted ammonium ions which are water-soluble. Carboxylic acids are organic acids that react with bases to form carboxylate ions which are water-soluble.

$$R-\underset{OH}{\overset{O}{\underset{\|}{C}}} + OH^- \rightleftharpoons R-\underset{O^-}{\overset{O}{\underset{\|}{C}}} + H_2O$$

$$\underbrace{R-NH_2}_{\substack{\text{Uncharged species} \\ \text{insoluble in water,} \\ \text{soluble in organic} \\ \text{solvents}}} + H^+ \rightleftharpoons \underbrace{R-NH_3^+}_{\substack{\text{Water-soluble} \\ \text{organic ions,} \\ \text{insoluble in organic} \\ \text{solvents}}}$$

Unlike the majority of organic functional groups which do not display acid/base reactions, carboxylic acids and amines can be extracted into water, providing the water has the appropriate pH. This property can be used to rapidly isolate either carboxylic acids or amines from other compounds which do not contain acidic or basic functional groups. In a more carefully controlled extraction, it is possible to accurately control the pH of an aqueous solution to selectively extract amines or acids, depending on the pK_a or pK_b of the acidic or basic functional groups that they contain. In this way, the components of a mixture containing a number of amines with different pK_b values can be separated.

The most general method for separating a mixture of organic compounds into its components is **chromatography**. This method separates organic compounds by making use of the ability of different organic compounds to be adsorbed to different extents onto a solid support. There is a wide variety of different solids that are used for chromatography and these solids include silica (SiO_2), alumina (Al_2O_3), synthetic and natural polymeric materials, and paper. The factors which influence the strength of adsorption include the polarity of groups on the solid surface (particularly the presence of —OH groups) the surface area available on the solid support, the shape and volume of cavities on the solid surface, as well as the presence or absence of cationic and anionic groups on the solid surface. Different organic compounds have a different

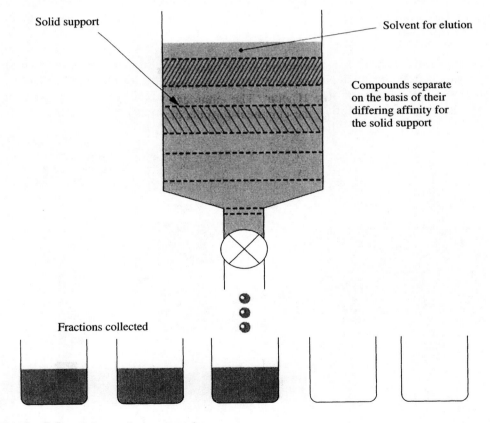

Figure 3.3 Column chromatography

affinity for a solid support and chromatography separates organic compounds on the basis of the affinity differences.

In the research laboratory, **column chromatography** (Figure 3.3) is commonly used. In this technique, a solid support is packed into a column and the mixture to be separated is adsorbed at the upper end. A solvent or solvent mixture is passed through the column (either percolating under gravity, or under pressure) to elute the adsorbed organic compounds. The solvent is collected in fractions after it emerges from the column. Components that are weakly adsorbed wash through the column rapidly and appear in the early fractions; conversely compounds that are strongly adsorbed take longer to travel the length of the column and appear in later fractions. There are many variations on column chromatography including **gas chromatography** (GC) and **high pressure liquid chromatography** (HPLC).

3.2 IDENTIFICATION OF ORGANIC COMPOUNDS

Once a compound has been isolated and purified, it is necessary to identify it. Firstly, it is desirable to obtain as much information as possible about its elemental composition and the functional groups the compound contains. A number of instrumental techniques are used to do this.

Elemental analysis

The empirical formula of a compound is the ratio of the different types of atoms in the compound. For example, benzene (C_6H_6) has the empirical formula of C_1H_1 because for each C atom there is one H atom. The empirical formula of acetylene (ethyne, C_2H_2) is also C_1H_1 so clearly, the empirical formula alone cannot be used to identify

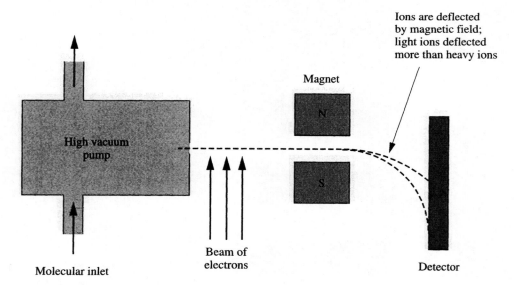

Figure 3.4 Schematic view of mass spectrometer

a compound, but it can be used to rule out many, if not most, possibilities. To determine the empirical formula it is necessary to measure how much of each element is present. This is done by *elemental analysis*. For many organic compounds it is sufficient to determine the amount of carbon, hydrogen, and nitrogen in the substance. If oxygen is the only other type of element present, the oxygen content can be calculated by difference once the percentage content of carbon, hydrogen, and nitrogen has been determined. In the analysis, a small, accurately weighed quantity of an unknown substance is burnt in a stream of pure oxygen. The carbon is quantatively converted to carbon dioxide, the hydrogen to water vapour, and the nitrogen present to nitrogen gas. The amount of each gas produced is determined, using a gas chromatograph. Modern elemental analysis is carried out using automated equipment.

Mass spectrometry

Mass spectrometry is a method used to determine the molecular weight of an organic compound. A small sample of the compound is vaporised under very low pressure and high temperature, and the vapour is irradiated with a beam of high energy electrons (Figure 3.4).

The effect of bombardment with the electron beam is to cause electrons to be ejected from molecules in the sample, leaving them as positively charged ions. The cations are accelerated through an electric field towards a negatively charged electrode and, travelling at high speed, they are released into a magnetic field. Because they are charged, the ions are deflected by the magnetic field. The lighter the ion, the stronger

Table 3.1 Common fragments and their masses

Fragment	Mass	Fragment	Mass
CH_3^+	15	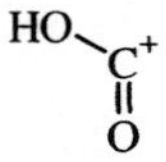 $HO-C^+(=O)$	45
$CH_3CH_2^+$	29	NO_2^+	46
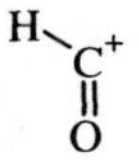 $H-C^+(=O)$	29	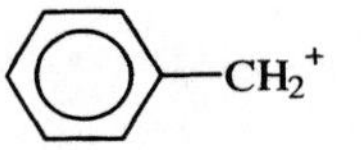 $C_6H_5-CH_2^+$	91
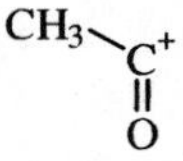 $CH_3-C^+(=O)$	43	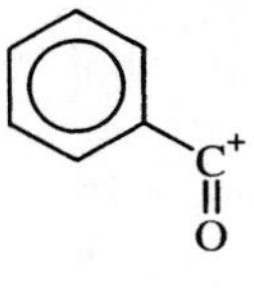 $C_6H_5-C^+(=O)$	105

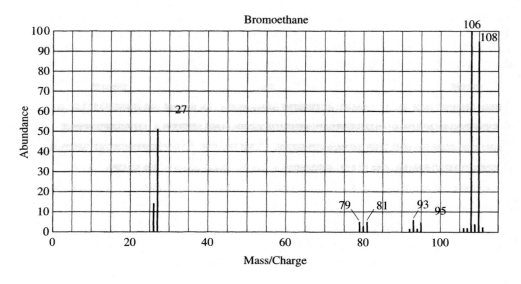
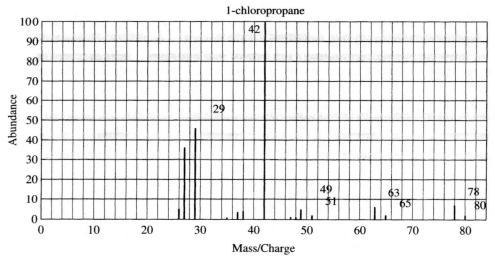

Figure 3.5 Mass spectra of bromoethane and 1–chloropropane

it is deflected by the magnetic field so the ions emerge from the magnetic field with different deflections, depending on their mass and charge. In principle, ions with more than a single charge are possible, but in practice these are rare. The ions are characterised by their mass to charge ratio (m/z).

A *mass spectrum* is a chart or graph of m/z against intensity. The spectrum of a compound typically shows a number of signals in the mass spectrum and the peak at highest m/z (termed the **parent ion** or *molecular ion*) usually corresponds to the mass of the whole molecule. The signals with lower m/z are called **daughter ions** or *fragment ions* and result from fragmentation of the ion before it enters the magnetic field. The fragment ions are useful because they identify discrete sections which can be cleaved from the parent ion. Some fragmentations are characteristic of a functional group. For example the bond adjacent to a carbonyl group breaks relatively easily to form an acyl cation (Table 3.1). The intensity of the signal is related to the stability of the fragment it represents and the strength of the bond broken in forming the fragment. The most intense fragment ions usually correspond to the most stable fragment ions.

A mass spectrum detects the mass of individual ion fragments, and each atom in a molecule could be one of several isotopes of a given element. Each fragment recorded in the mass spectrum not only registers the elements present in the molecule, but also specific isotopes of the various elements present. For some elements there is more than one isotope of high natural abundance (for example, bromine has isotopic abundances—^{79}Br 49% and ^{81}Br 51%; chlorine has isotopic abundances—^{35}Cl 75% and ^{37}Cl 25%). This means that for any ion which contains a bromine atom, the mass spectrum contains two signals separated by two m/z units, one for the ions which contain ^{79}Br and one for the ions which contain ^{81}Br. For bromine-containing ions, the relative intensities of the two signals are approximately the same since the natural abundances of ^{79}Br and ^{81}Br are approximately the same (Figure 3.5). For any ion which contains a chlorine atom, the mass spectrum contains two signals separated by two m/z units, one for the ions which contain ^{35}Cl and one for the ions which contain ^{37}Cl. In this case the relative intensities of the two signals are approximately 3:1 since this reflects the natural abundances of ^{35}Cl and ^{37}Cl (Figure 3.5).

In many instances it is not possible to assign a molecular formula to a compound on the basis of the m/z of its parent ion. For example, a parent ion at m/z 72 could be due to a compound whose molecular formula is C_4H_8O or one with a molecular formula $C_3H_4O_2$ or one with a molecular formula $C_3H_8N_2$. If the mass spectrum is recorded with extremely high precision (*a high resolution mass spectrum*) then the mass of the parent ion, or any fragment, can be recorded to much better than integer precision. Since the mass of the atoms of each element is known to high accuracy, molecules which have the same mass when measured only to the nearest integer mass unit, can usually be distinguished when the mass is measured with high precision. The accurate masses of ^{12}C, ^{16}O, ^{14}N and ^{1}H are 12.0000 (by definition), 15.9949, 14.0031, and 1.0078, so ions with the formulas $C_4H_8O^+$, $C_3H_4O_2^+$ or $C_3H_8N_2^+$ would have masses 72.0573, 72.0210, and 72.0686 respectively, which could easily be distinguished by high resolution mass spectrometry. Additionally, if the mass of any fragment in the mass spectrum can be accurately determined, then there is usually only one combination of elements which can give rise to that signal since there are only a limited number of elements and their masses are accurately known.

SAMPLE EXERCISE 3.1

The parent ion in the mass spectrum of an unknown compound occurs at m/z 78.0105 when it is recorded at high resolution. Would this correspond to an ion with the formula $C_6H_6^+$, $C_5H_2O^+$ or $C_5H_4N^+$?

Solution: From the high resolution masses given above, $C_6H_6^+$, $C_5H_2O^+$ and $C_5H_4N^+$ would have masses 78.0468, 78.0105 and 78.0343. Clearly, the mass of 78.0105 corresponds to an ion with the formula $C_5H_2O^+$.

Catalogues (and now electronic databases) of the mass spectral fragmentation patterns of known molecules are now available and can be rapidly searched by computer. The pattern and intensity of fragments in the mass spectrum is characteristic of an individual compound so comparison of the experimental mass spectrum of a compound with those in a library can be used to positively identify the compound, if its spectrum has been recorded. It is now quite common to couple an instrument for separating a mixture of organic compounds, for example a gas chromatograph, directly to the input of a mass spectrometer. In this way as each individual compound is separated from the mixture, its mass spectrum can be recorded and compared to the library of known compounds and identified immediately if it is a known compound.

SAMPLE EXERCISE 3.2

Which of the following compounds would have parent ions in their mass spectrum at m/z of 60?

(a) $CH_3CH_2—O—CH_3$

(b) $CH_3CH_2CH_2—OH$

(c) $CH_3—CH(OH)—CH_3$

(d) 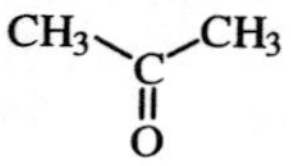 $CH_3—C(=O)—CH_3$

(e) 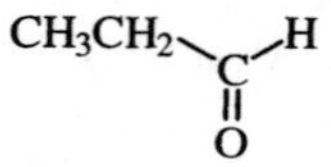 $CH_3CH_2—C(=O)—H$

Solution: Compounds (a), (b) and (c) are constitutional isomers with formula C_3H_8O and molar mass 60 g mol^{-1}. Compounds (d) and (e) have the formula C_3H_6O and a molar mass of 58 g mol^{-1}.

SAMPLE EXERCISE 3.3

Which of the following compounds could give rise to fragment ions at m/z of 43 in their mass spectrum?

(a) 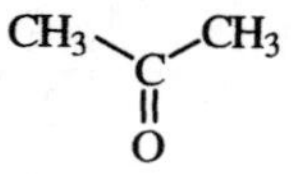 $CH_3—C(=O)—CH_3$

(b) $CH_3—C(=O)—O—CH_2CH_3$

(c) phenyl—C(=O)—CH$_3$

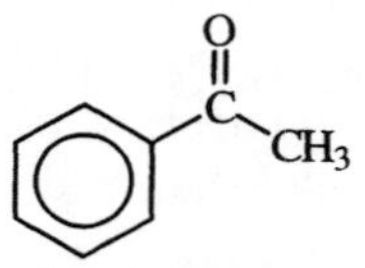

(d) $CH_3CH_2—O—CH_2CH_3$

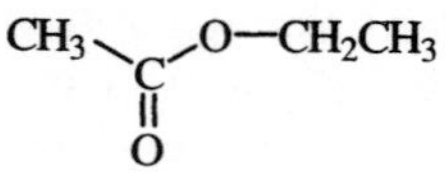

(e) 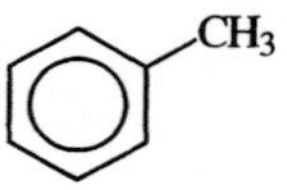 phenyl—CH$_3$

Solution: Compounds (a), (b) and (c) can all give rise to the $[CH_3C{=}O]^+$ fragment which has m/z of 43.

PRACTICE EXERCISE

Which of the following compounds would have the parent ion in their mass spectrum at m/z of 72 ?

(a) $CH_3CH_2-\underset{\underset{O}{\|}}{C}-CH_3$ (b) $CH_3CH_2CH_2-\underset{\underset{O}{\|}}{C}-H$ (c) phenyl-Cl

(d) CH_3Cl (e) tetrahydrofuran

Answer: (a), (b), and (e).

Absorption spectroscopy

Organic compounds interact and absorb radiation in different regions of the electromagnetic spectrum. All electromagnetic radiation is characterised by the fact that it travels with the same velocity (the velocity of light), (c) and its wavelength (λ) and its frequency (ν) are related by the equation $\nu = c/\lambda$. Since c is constant ($ca\ 3 \times 10^8$ m sec^{-1}), this equation indicates that the longer the wavelength of the radiation, the smaller its frequency. The energy of radiation is proportional to its frequency so radiation with low frequency (and long wavelength) is comparatively low energy radiation ($E = h\nu$) and conversely radiation with high frequency (and short wavelength) is comparatively high energy radiation. In the electromagnetic spectrum, radio waves are at the low energy end and gamma rays and cosmic rays have the highest energy (Table 3.2).

Organic compounds absorb electromagnetic radiation in three major regions of the electromagnetic spectrum, (i) the UV-visible region (ii) the IR region (iii) the Rf region.

Table 3.2 Wavelengths and frequencies of electromagnetic radiation

Radiation	Approximate wavelength (m)	Approximate frequency (Hz)
Cosmic rays	1×10^{-14}	3×10^{22}
Gamma rays	1×10^{-12}	3×10^{20}
X-rays	1×10^{-9}	3×10^{17}
Ultraviolet light	1×10^{-8}	3×10^{16}
Visible light	5×10^{-7}	6×10^{14}
Infrared light	3×10^{-6}	1×10^{14}
Microwaves	3×10^{-2}	1×10^{10}
Television waves	3	1×10^{8}
Radio waves	3×10^{1}	1×10^{7}

Increasing energy ↑

Radiation in each of these regions of the spectrum has the correct energy to interact with specific parts of an organic molecule. The absorption of electromagnetic radiation can be detected and used to identify features of the molecule and this is termed **absorption spectroscopy**.

Ultraviolet-visible spectroscopy

Radiation in the ultraviolet (UV) and visible region of the spectrum has the correct energy to excite electrons in one orbital into an orbital of higher energy. The electrons which are most easily promoted are those in **conjugated** π-bonds. A conjugated molecule is one in which there is an alternation between single and multiple bonds in at least part of the molecule (Figure 3.6). Conjugated molecules include aromatic compounds, 1,3–dienes, (for example, $H_2C=CH-CH=CH_2$), 1,3–diynes, (for example, $HC\equiv C-C\equiv CH$) and α,β-unsaturated carbonyl compounds (for example, propenal, $H_2C=CH-CHO$).

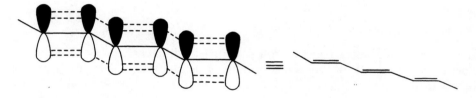

In a conjugated system of double or triple bonds, alteration between single and multiple bonds permits interaction between adjacent π-systems

Figure 3.6 Conjugation in a triene molecule.

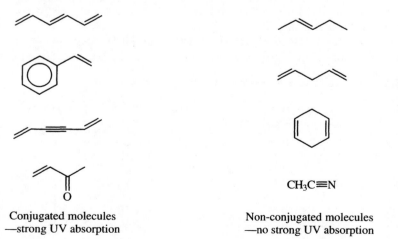

Conjugated molecules
—strong UV absorption

Non-conjugated molecules
—no strong UV absorption

Figure 3.7 Examples of molecules containing a π-system

Organic compounds which contain conjugated multiple bonds strongly absorb UV or visible radiation, (Figure 3.7) and may be identified using **UV-visible spectroscopy**.

An UV-visible spectrum is a graph of absorption intensity against the wavelength of absorption. The UV range extends from 200 to 400 nm and the visible range is from 400 to 800 nm. A typical UV-visible absorption spectrum is broad and lacks sharp features, for example, Figure 3.8. In its simplest sense, the presence of a strong absorption in the UV-visible region of the spectrum indicates the presence of conjugation in an organic molecule. Because different compounds have different bonding arrangements and hence different electron energy levels, the UV-visible spectrum of a compound is a unique characteristic of the compound.

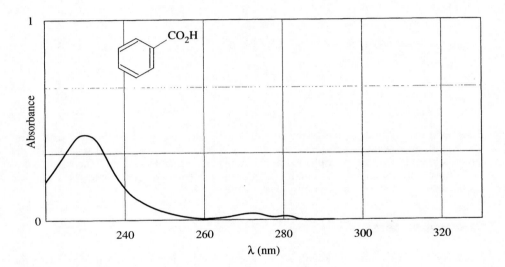

Figure 3.8 The UV spectrum of benzoic acid

SAMPLE QUESTION

Which of the following compounds strongly absorb electronic radiation in the UV region of the spectrum?

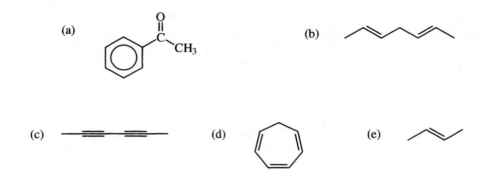

Solution: (a), (c) and (d) are conjugated molecules and strongly absorb UV radiation; in (a) the carbonyl bond is separated by one single bond from the π-bonded atoms in the aromatic ring, in (c) and (d) triple and double bonds are separated by only one single bond. Molecules (b) and (e) are not conjugated because in (b) two single bonds separate the double bonds and in (e) there is only one double bond.

PRACTICE PROBLEM

Which of the following compounds strongly absorb electronic radiation in the UV region of the spectrum?

Answer: (b), (c) and (e).

Table 3.3 Characteristic infrared absorption frequencies of some bonds

Bonds	Region of IR absorption (cm^{-1})	Bonds	Region of IR absorption (cm^{-1})
C—H (alkanes)	2850–2950	C≡C	2100–2300
N—H	3200–3600	C≡N	~ 2250
O—H (alcohol)	3500–3650 (broad)		
O—H (COOH)	2500–3600 (broad)	C=O	1650–1850 (strong)

Infrared spectroscopy

Electromagnetic radiation in the infrared (IR) region of the spectrum has the correct energy to cause bonds in a molecule to stretch and angles to bend. The energy required to stretch a bond depends on the mass of the bonded atoms and the strength of the bond between them. In general, the energy required to stretch the bond between light atoms is higher than the energy required for a bond between heavier atoms. Stronger bonds (for example, double bonds and triple bonds) absorb infrared radiation of higher energy than weaker bonds (for example, single bonds). This is the basis of **Infrared spectroscopy**.

An IR spectrum is a graph of IR absorption intensity against the frequency of

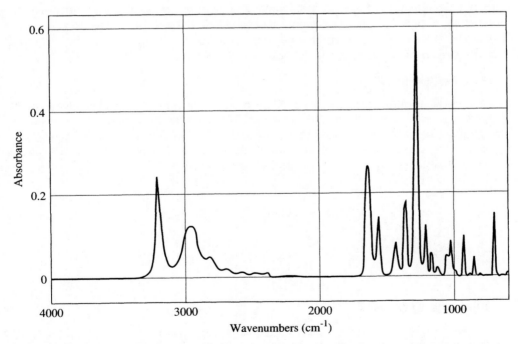

Figure 3.9 An IR spectrum of a compound containing C=O, N—H and C—H bonds

absorption. The IR range extends from about 400 to 4000 cm^{-1} (1 cm^{-1} = 3.0 × 10^{10} Hz). Individual functional groups have characteristic regions of the IR spectrum where they have absorption bands and this means that an IR spectrum is most useful in establishing the presence or absence of specific functional groups in a molecule (Table 3.3). For example, the C=O group of aldehydes, ketones, esters, carboxylic acids, and so on, strongly absorbs IR radiation near 1700 cm^{-1}. The —NH group of amines, and primary and secondary amides, strongly absorbs IR radiation in the range 3200 to 3600 cm^{-1}, (Figure 3.9). The absence of an absorption can also be important. For example, if an oxygen-containing compound shows no absorption in the C=O region (1680 to 1750 cm^{-1}) or in the O—H region (2500 to 3650 cm^{-1}) of the IR spectrum, the compound is likely to be an ether.

In practice, an IR spectrum is obtained by passing a beam of IR radiation through a thin sample of the material and monitoring the intensity of the beam as a function of the frequency of the radiation. The sample for IR spectroscopy is typically a solid, a solution, or a liquid film.

The region of an IR spectrum below about 1500 cm^{-1} is termed the **fingerprint region**. Many absorptions in this region of the spectrum are the result of complex composite vibrations of the molecule as a whole or sections of the molecule. These vibrations depend on the specific arrangement of bonds, atoms, and functional groups in the molecule and no two molecules have exactly the same absorption in the fingerprint region. The IR absorption spectrum in the fingerprint region is a diagnostic vibrational signature of an organic compound and serves to uniquely identify the compound. Tables (and now electronic databases) of IR absorption spectra serve as a permanent library of known compounds and an unknown compound can be identified by comparison of its spectrum with those in such a library.

SAMPLE EXERCISE 3.4

A compound with molecular formula C$_3$H$_6$O has a strong absorption at 1720 cm^{-1} in its IR spectrum. Give reasonable condensed structural formulas for this compound.

Solution: Strong absorption near 1720 cm^{-1} in its IR spectrum is indicative of the presence of a carbonyl group in the compound. With the molecular formula C$_3$H$_6$O the compound would have to be the ketone or the aldehyde shown here.

$$CH_3\text{-}\underset{\underset{O}{\|}}{C}\text{-}CH_3 \qquad CH_3CH_2\text{-}\underset{\underset{O}{\|}}{C}\text{-}H$$

PRACTICE EXERCISE

A compound with molecular formula C$_3$H$_8$O has a strong absorption at 3600 cm^{-1} in its IR spectrum. Give reasonable condensed structural formulas for compounds whose structures are consistent with this data.

Answer:

$$CH_3CH_2CH_2OH \qquad CH_3\underset{\underset{OH}{|}}{C}HCH_3$$

Nuclear Magnetic Resonance (NMR) spectroscopy

The nuclei of some isotopes of many elements absorb electromagnetic radiation in the radiofrequency (Rf) region of the spectrum when they are placed in a strong magnetic field. The measurement of the absorption of Rf radiation by nuclei in a magnetic field is called **Nuclear Magnetic Resonance spectroscopy** and the name is commonly abbreviated to NMR spectroscopy or simply MR spectroscopy. An NMR spectrometer is an instrument designed to detect and record the frequencies (therefore energies) of Rf energy absorbed by the various nuclei in an organic compound. An NMR spectrometer consists of a large magnet with an intense magnetic field, a transmitter to provide the Rf energy, a receiver to detect the absorption of Rf radiation by the sample and a recording device to register and display the NMR spectrum (Figure 3.10).

Of the elements commonly encountered in organic chemistry, the nuclei of hydro-

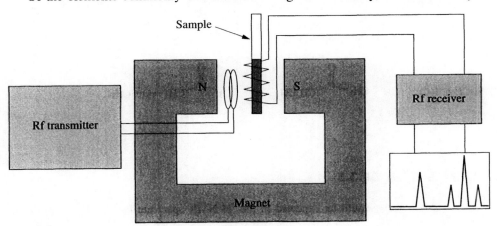

Figure 3.10 Schematic of an NMR spectrometer

gen atoms (^{1}H) give rise to the most sensitive and widely used form of NMR spectroscopy (^{1}H NMR). NMR can be used to observe other elements in the periodic table including ^{31}P, ^{19}F, and ^{13}C. ^{13}C NMR is important in analysing organic molecules, however ^{13}C is *not* the most abundant isotope of carbon (^{13}C is about 1.1% of all carbon) and this makes ^{13}C NMR far less sensitive than ^{1}H NMR spectroscopy.

The nuclei of all chemically equivalent hydrogen atoms in a molecule absorb Rf radiation at the same frequency or, conversely, hydrogen atoms in different chemical environments within a molecule absorb Rf radiation at different frequencies. For example, methyl acetate has two distinct hydrogen environments; the hydrogen atoms of the methyl group directly attached to the oxygen atom are in a different chemical environment to the hydrogen atoms of the methyl group bonded to the carbonyl carbon atom (Figure 3.11). The two different hydrogen environments absorb Rf radiation of different frequences. The number of signals in an NMR spectrum corresponds to the number of distinct types of hydrogen atoms in a molecule and this is determined by the symmetry of the molecule.

One signal in the ^{1}H NMR spectrum

CH_4 CH_3-CH_3 CH_3-O-CH_3 $CH_3-\overset{\overset{O}{\|}}{C}-CH_3$ cyclopentane

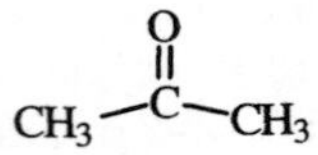

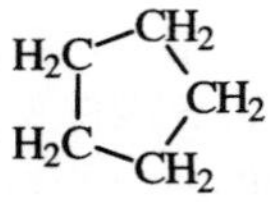

Two signals in the ^{1}H NMR spectrum

CH_3CH_2Cl $CH_3CH_2-O-CH_2CH_3$ $CH_3-\overset{\overset{O}{\|}}{C}-O-CH_3$

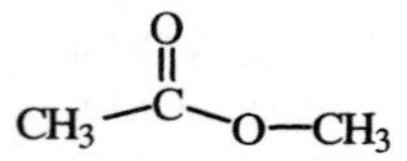

Three signals in the ^{1}H NMR spectrum

$CH_3CH_2CH_2Cl$ $CH_3-\overset{\overset{O}{\|}}{C}-CH_2CH_3$ $CH_3-\overset{\overset{O}{\|}}{C}-CH_2-O-CH_3$

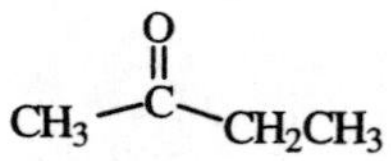

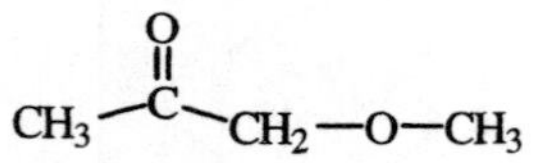

SAMPLE EXERCISE 3.5

How many signals would be expected in the ^{1}H NMR spectrum of each of the following compounds:

(a) CH_3-Br (b) $CH_3-\underset{\underset{CH_3}{|}}{\overset{\overset{CH_3}{|}}{C}}-CH_3$ (c) $CH_3CH_2CH_3$

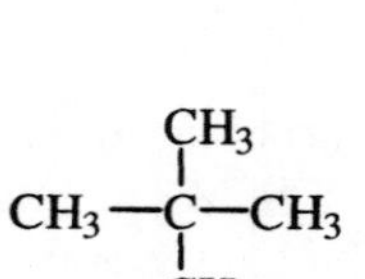

Solution: In (a), the carbon atom of the methyl group is sp^3-hybridised so the hydrogen atoms are at the corners of a tetrahedron. All of the hydrogen atoms in a CH_3 group are equivalent and there would be only one signal in this ^1H NMR spectrum. In (b), the central carbon atom is sp^3 hybridised and this means that the four methyl groups are attached at the corners of a tetrahedron. All of the CH_3 groups would be identical to each other and so there would be only one signal in the ^1H NMR spectrum of (b). In (c) the molecule has a plane of symmetry through the central carbon atom so the CH_3 groups are identical. The ^1H NMR spectrum of (c) would have two signals—one from the two hydrogen atoms on the central carbon atom and one for the six hydrogen atoms of the equivalent methyl groups.

PRACTICE EXERCISE

How many signals would be expected in the ^1H NMR spectrum of each of the following compounds:

(a) $CH_3-\underset{\underset{H}{|}}{\overset{\overset{CH_3}{|}}{C}}-CH_3$

(b) $CH_3CH_2-\overset{\overset{O}{\|}}{C}-CH_2CH_3$

(c) $Br-CH_2CH_2CH_2-Br$

(d) $ClCH_2-\overset{\overset{O}{\|}}{C}-CH_2Cl$

Answers: (a) 2; (b) 2; (c) 2; (d) 1

An NMR spectrum is a graph of the frequency of absorption against Rf absorption intensity. The frequency of absorbed radiation depends on the strength of the magnetic field as well as the environment of the absorbing nucleus. The frequency axis of the spectrum is calibrated with a scale (called the δ-scale) in dimensionless units called

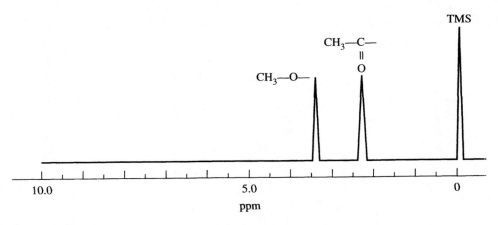

Figure 3.11 Schematic representation of the ^1H NMR spectrum of methyl acetate

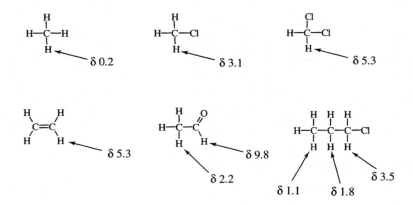

Figure 3.12 Representative ¹H NMR chemical shifts in organic molecules

parts per million (ppm). The δ-scale is not in absolute frequency units—the units are 'normalised' so that the ppm scale gives the same numbers irrespective of the strength of the magnet in which the measurement was made. The δ-scale is calibrated with respect to the signal of a reference compound added to the sample. For ¹H NMR, the reference compound is tetramethylsilane $(CH_3)_4Si$ (abbreviated as TMS). TMS gives a sharp signal, outside the range of normal organic compounds and its frequency is always taken as δ = 0.00 ppm The frequency difference between a signal arising from a sample and the TMS signal is called the **chemical shift**.

The chemical shifts of all signals in an NMR spectrum are measured in ppm (δ scale) from the TMS signal. The ¹H NMR signals from hydrogen atoms in alkyl groups come typically in the region 0 ppm to 2 ppm from TMS, the hydrogen atoms attached to carbon atoms adjacent to carbonyl groups typically occur 2 ppm to 3 ppm from TMS and the hydrogen atoms attached to aromatic rings occur in the region 7 ppm to 8 ppm (Figure 3.12).

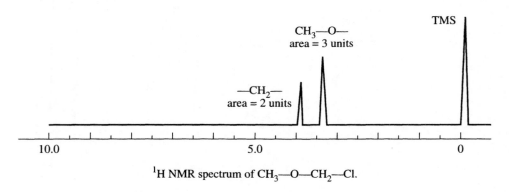

Figure 3.13 Schematic representation of the ¹H NMR spectrum of chloromethyl methyl ether

Table 3.4 Signal multiplicities in ¹H NMR spectra

number of neighbouring hydrogen atoms (n)	signal multiplicity	name of signal
0	1	singlet
1	2	doublet
2	3	triplet
3	4	quartet
4	5	quintet
.	.	...

The hydrogen atoms in most organic molecules give rise to signals in the range 0 ppm to 13 ppm from TMS. The *relative intensities* of the various signals in a ¹H NMR spectrum (as measured by the signal area) are proportional to the number of hydrogen atoms which give rise to the signal. The ¹H NMR spectrum of CH_3—O—CH_2—Cl would exhibit two signals (one for the hydrogen atoms which are part of the CH_3— group and one for the hydrogen atoms which are part of the —CH_2— group) and these would have intensities in the ratio 3:2 (Figure 3.13).

The signals in a ¹H NMR spectrum frequently show fine structure which is termed *splitting* or **multiplicity**. The splitting of a signal from a hydrogen atom or group of hydrogen atoms is due to the presence of other hydrogen nuclei attached to the adjacent carbon atom. The nuclei of hydrogen atoms behave as if they were magnetised and the splitting of an NMR signal arises from the fact that each magnetic nucleus can sense the presence of other magnetic nuclei nearby *through the bonds of the molecule*. The splitting pattern of an NMR signal in the ¹H NMR spectrum arises from the hydrogen atoms attached to the neighbouring carbon atoms. If a hydrogen atom has 'n' equivalent hydrogen atoms on adjacent carbon atoms, its NMR resonance will appear as a signal which is split into $n + 1$ lines—this is called the $n + 1$ *rule* (Table 3.4).

The NMR signal of a hydrogen nucleus is not split by other hydrogen atoms which are the same as itself—the NMR signal for ethane (CH_3—CH_3) is a singlet with no splitting because all of the hydrogen nuclei are the same and no hydrogen is split by any of the others. The signal of a hydrogen nucleus is generally only visibly split by other hydrogen nuclei which are *no more than three bonds away*. Although there can be splitting due to more remote hydrogens, the splitting effect diminishes rapidly as the number of bonds between the hydrogen atoms increases so that only the large splittings due to hydrogen atoms of immediate carbon neighbours can be seen easily.

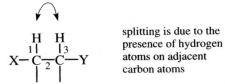

splitting is due to the presence of hydrogen atoms on adjacent carbon atoms

An ethyl group not adjacent to a hydrogen-bearing atom ('isolated') always gives a triplet signal (of relative intensity three units) for the CH_3 group and a quartet signal (of relative area two units) for the CH_2 group. The signal for the CH_3 group appears as a triplet because on the adjacent carbon atom there are two hydrogen atoms ($n + 1 = 2 + 1 = 3$). The quartet splitting of the CH_2 group arises because on the adjacent carbon there are three hydrogen atoms ($n + 1 = 3 + 1 = 4$) (Figure 3.14).

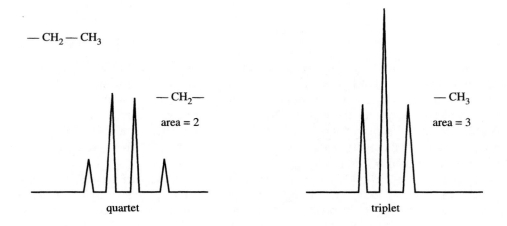

Figure 3.14 The splitting of the signals in an isolated ethyl group

Quartets characteristically have relative line intensities 1:3:3:1; triplets 1:2:1, and doublets are composed of two lines of equal intensity (Figure 3.15).

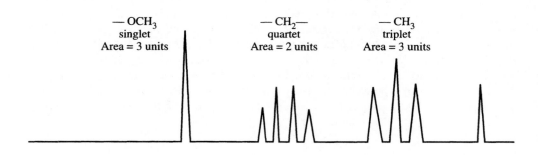

Figure 3.15 Schematic representation of the ^1H NMR spectrum of methyl propionate

SAMPLE EXERCISE 3.6

Give the number of signals, their relative intensities and the multiplicities in ^1H NMR spectra of each of the following compounds:

(a) CH₃CH₂Cl

(b) CH₃—C(CH₃)(H)—Cl

Solution: In (a) the CH$_2$ and CH$_3$ groups are different so there will be two signals. Three hydrogen atoms give rise to the signal from the CH$_3$ and two hydrogen atoms contribute to the signal for the CH$_2$ so the signals will have relative intensities 3:2. For the CH$_3$ group there are two hydrogen atoms on the adjacent carbon atom so the multiplicity of the signal of the CH$_3$ group will be three (a *triplet*). For the CH$_2$ group there are three hydrogen atoms on the adjacent carbon atom so the multiplicity of the signal of the CH$_2$ group will be four (a quartet). In (b) the two CH$_3$ groups are equivalent but different to the CH. There will be two signals. Six hydrogen atoms give rise to the signal from the CH$_3$ groups and one hydrogen atom contributes to the signal for the CH so the signals will have relative intensities 6:1. For the CH$_3$ groups there is only one hydrogen atom on the adjacent carbon atom so the multiplicity of the signal of the CH$_3$ groups will be two (a *doublet*). For the CH group there are six hydrogen atoms on the adjacent carbon atoms so the multiplicity of the signal of the CH group will be seven (a *septet*).

PRACTICE EXERCISE

Give the number of signals, their relative intensities and the multiplicities in ^{1}H NMR spectra of each of the following compounds:

(a) CH₃CH₂—O—CH₃

(b) CH₃—C(H)(CH₃)—O—CH₂CH₃

(c) CH₃—C(CH₃)(CH₃)—CH₂Br

Answers: (a) three signals with relative areas 3:2:3 and multiplicities triplet, quartet, and singlet, respectively; (b) four signals with relative areas 6:1:2:3 and multiplicities doublet, septet (seven lines), quartet and triplet, respectively; (c) two signals with relative areas 9:2 and both with singlet multiplicities.

Crystal structure analysis

All of the techniques for identifying compounds that have been discussed so far rely on some degree of interpretation and consequently there is the potential for a misinterpretation or error. It is not possible to look at, or photograph, molecules, even with the most powerful microscopes, because the separation of the atoms is smaller than the wavelength of visible light. However, by using radiation of wavelength close to that of the interatomic separation it is possible to produce a 'picture of a molecule'. Most commonly, this radiation is X-ray radiation, but it can also be a stream of high energy electrons or neutrons and the technique used to produce the pictures is called *crystal structure analysis*. As the name suggests, this technique requires the compound under study to be in the crystalline state. If a good quality crystal of a compound is

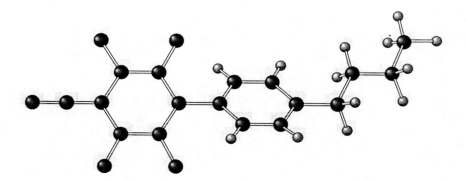

Figure 3.16 The structure of 4-butyl-4′-cyano-2′, 3′, 5′, 6′-tetrafluorobiphenyl as determined by x-ray crystal structure analysis

available then using modern equipment it is possible to 'determine the structure of the compound' in a few days or less.

When radiation of the types mentioned above strikes a crystal the radiation is diffracted. Depending on the angle the radiation strikes the crystal it encounters a different sequence of atoms and the intensity of the diffracted beam is differently affected. Crystal structure analysis relies on taking a series (usually thousands) of these intensities and back calculating, by means of a Fourier transformation, what atom positions must have caused the observed distribution of intensities. Once the atom positions are known it is easy to work out which atoms are connected and so determine what the molecule is. An example of a picture of a compound determined by crystal structure analysis is shown in Figure 3.16. Detailed information, such as bond lengths and angles, is obtained for a molecule using this technique.

3.3 THE THREE-DIMENSIONAL STRUCTURE OF ORGANIC COMPOUNDS —STEREOCHEMISTRY OF ORGANIC COMPOUNDS

Given the molecular formula for an organic compound, there are still a number of possible structures that the formula represents. There is usually a number of constitutional isomers for any molecular formula—recall that constitutional isomers differ by the nature and sequence of atoms and bonds. Also, for each of the constitutional isomers there may be **stereoisomers**—stereoisomers have the same sequence of atoms and bonds, but differ in the arrangement of the atoms in space. Alkenes can have *E* or *Z* stereoisomers depending on the relative arrangement of groups attached to the C=C double bond (Section 1.2). Substituted cyclic compounds can have *cis* or *trans* isomers depending on whether the substituents are attached on the same or opposite faces of the ring.

The existence of isomerism in organic compounds stems from the fact that carbon atoms which are sp^3-, sp^2-, or sp-hybridised have characteristic bonding geometries and once bonds are formed it is difficult to change the relative positions of the bonded

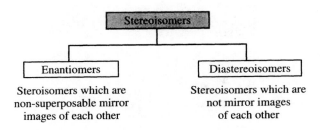

Figure 3.17 Classes of stereoisomers

atoms and groups without breaking and re-forming bonds. To do this requires considerable energy so most stereoisomers of a compound are stable and do not interconvert at room temperature. They frequently possess different physical and chemical properties to one another.

Some molecules also exhibit **chirality** or *optical isomerism* which arises when a molecule and its mirror image are not identical (not superposable). A pair of molecules which are not identical, but are mirror images of each other, are called **enantiomers**. Enantiomers are still stereoisomers. However, in contrast to the other types of stereoisomers, almost all of the properties of enantiomers are identical (melting point, boiling point, solubility, etc.). All stereoisomers which are not enantiomers are grouped together in a broad category called **diastereoisomers**, (Figure 3.17).

Optical isomerism

The bonding orbitals of an sp^3-hybridised carbon atom point towards the corners of a tetrahedron. The angles between the bonded groups are approximately 109.5° and the three-dimensional arrangement of the bonds around the carbon atom can be represented pictorially as two lines of normal thickness (bonds in the plane of the paper), one dashed line (a bond which points away from the observer, behind the plane of the paper) and a thickened or wedge line (a bond which points forward, towards the observer). This notation is commonly used to draw the structure of a molecule, or part of a molecule, where it is important to indicate the three-dimensional arrangement of the groups (Figure 3.18).

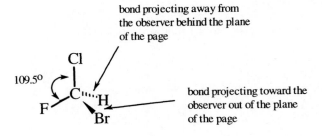

Figure 3.18 Three-dimensional representation of an sp^3-hybridised carbon atom

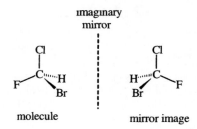

Figure 3.19 A chiral compound and its mirror image

The mirror image of any molecule can be constructed by drawing an imaginary mirror and constructing the molecule which would be observed if it were viewed in the mirror.

Identical molecules are superposable—when superposable molecules, groups or objects are placed on top of each other, all parts coincide. An sp^3-hybridised carbon atom which has four different groups attached to it is not identical to, or non-superposable on, its mirror image. Such a carbon atom is called a **stereogenic centre** or a **chiral carbon centre**. If the mirror image molecule of bromochlorofluoromethane (Figure 3.19) is rotated so that the Cl and F groups coincide with each other, the other two groups do not coincide (Figure 3.20).

A molecule and its mirror image are not superposable when there is a stereogenic centre present

Figure 3.20 Non-superposition of a molecule containing a stereogenic centre and its mirror image

SAMPLE EXERCISE 3.7

Identify any stereogenic centres in the following compounds:

(a)
```
    H H Cl H
    | | | |
H—C—C—C—C—H
    | | | |
    H H H H
```

(b)
```
    H H H CH₃
    | | | |
H—C—C—C—C—CH₃
    | | | |
    H OH H H
```

Solution: Any carbon atom which has two identical substituents cannot be a stereogenic centre. In (a) the carbon identified with an asterisk (*) is a stereogenic centre—it has a —Cl, an —H, a —CH_3 and a —CH_2CH_3 group attached to it. In (b) the carbon identified with an asterisk (*) is a stereogenic centre—it has an —OH, an —H, a —CH_3 and a —$CH_2CH(CH_3)_2$ group attached to it.

(a) 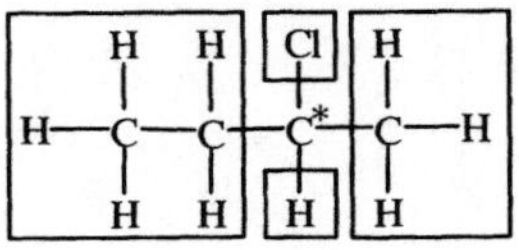 H—C(H)(H)—C(H)(H)—C*(Cl)(H)—C(H)(H)—H

(b) 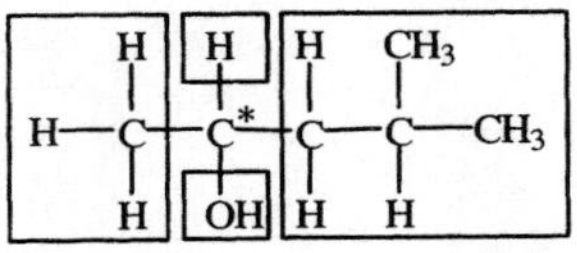 H—C(H)(H)—C*(H)(OH)—C(H)(H)—C(H)(CH_3)—CH_3

PRACTICE EXERCISE

Identify any stereogenic carbon centre in each of the following compounds:

(a) 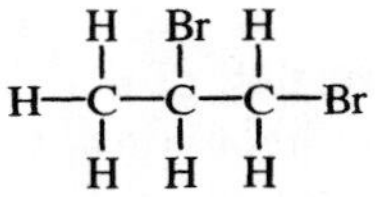 H—C(H)(H)—C(Br)(H)—C(H)(H)—Br

(b) 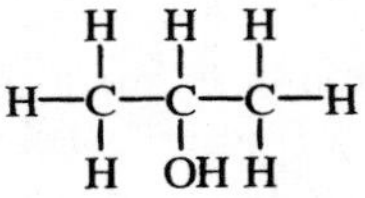 H—C(H)(H)—C(H)(OH)—C(H)(H)—H

Answers:

(a) 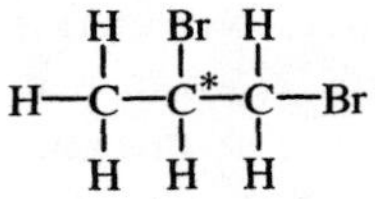 H—C(H)(H)—C*(Br)(H)—C(H)(H)—Br

(b) There are no stereogenic centres

SAMPLE EXERCISE 3.8

Using wedge and dotted lines to represent bonds which are out of the plane of the paper, draw the two enantiomers of 2–butanol.

Solution: In 2–butanol, $CH_3CH(OH)CH_2CH_3$, the stereogenic centre has an —OH, an —H, a —CH_3 and a —CH_2CH_3 group attached to it. Construct one of the enantiomers by first drawing the carbon atom at the stereogenic centre and attaching the four different substituents to it. Then draw an imaginary mirror plane and construct the second molecule as the image of the first would appear in the mirror (Figure 3.21).

Enantiomers differ only in (i) the way that they react or interact with other chiral molecules, (ii) sometimes in the shape of the crystals that they form, and (iii) the way that they interact with polarised light. The ability of enantiomers to interact differently with polarised light is important since it provides a convenient and reliable method for characterising them. If the members of a pair of enantiomers can be isolated in a pure form, one will *rotate the plane of polarised light* in a given direction and the other

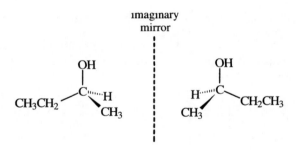

Figure 3.21 Two enantiomers of 2–butanol

enantiomer will rotate the plane of light of the same frequency in the opposite direction by the same amount. Molecules which rotate plane-polarised light are called **optically active** molecules.

The polarimeter

A **polarimeter** is an instrument used to detect the rotation of the plane of polarisation of a beam of polarised light. A beam of normal white light consists of waves vibrating in all planes perpendicular to the direction of travel. Polarised light consists of waves vibrating only in one of the planes perpendicular to the direction of travel. Polarised light is formed by passing a beam of light through a polarising filter which is constructed of special material which only transmits light waves oscillating in one direction. Light can pass through a series of polarising filters provided they are aligned such that the planes of the light that they transmit are parallel. However if two polarising filters are aligned such that the planes of the light that they transmit are at right angles, a light beam polarised by the first filter cannot pass through the second filter. No light emerges from a pair of crossed polarising filters.

A polarimeter consists of a pair of orthogonal polarising filters with a sample compartment between them. The first polarising filter generates a beam of polarised light; the second polarising filter is used as an analyser which can be rotated to measure accurately the plane of polarisation of the light. In a modern instrument the measurement is automated. The two polarising filters are initially set so as not to transmit any light. After the test sample is introduced, the analysing filter is rotated by an amount necessary to null the transmitted light (Figure 3.22).

The angle through which a sample rotates the plane of polarised light is called *the observed rotation* and it is expressed in degrees. The rotation depends on the nature of the sample, its enantiomeric purity, its concentration, the size of the polarimeter cell containing the sample solution, and the wavelength of light. If the rotation is recorded under standard conditions with a concentration of 1 g mL^{-1} for a sample container 10 cm long, the observed rotation is given the symbol [α] and called the **specific rotation** of the molecule. Samples which require the analysing filter to be rotated in a clockwise direction are by convention given a positive (+) sign. Conversely, samples which require the analysing filter to be rotated in an anti-clockwise direction have a negative (-) sign.

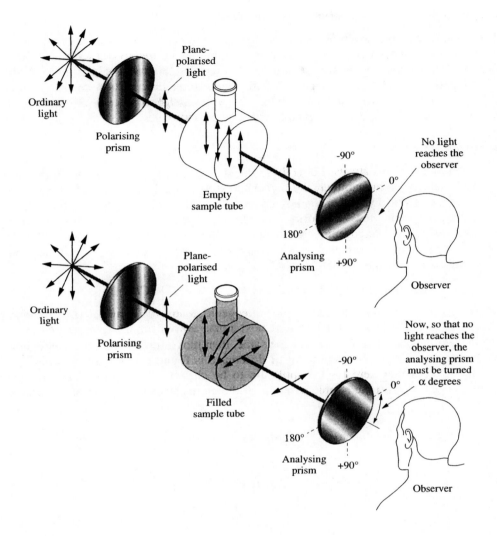

Figure 3.22 Schematic diagram of a polarimeter

If a pair of enantiomers is isolated in a pure state, one of the enantiomers will rotate the plane of polarised light of a particular frequency in a clockwise direction and the other in an anti-clockwise direction. For example, if one enantiomer has an [α] of +20°, the other will have an [α] of −20°. It is not currently possible to predict the direction or the magnitude of optical rotation from the molecular structure, however if one member of a pair of enantiomers has been isolated and its rotation measured, we can predict that the other enantiomer must cause a rotation which is opposite in sign, but equal in magnitude.

The absolute configuration at a stereogenic centre

The **absolute configuration** of a stereogenic centre refers to the exact three-dimensional arrangement of the groups. The method for naming enantiomers is called the *Cahn–Ingold–Prelog convention* and is based on describing the relative positions of the substituents attached to a stereogenic centre. Each stereogenic carbon centre has four different groups attached and these must firstly be assigned a relative sequence or priority according to a set of **sequence rules** or priority rules.

The sequence rules

(i) Each of the atoms attached directly to the stereogenic centre is assigned a priority, based on its atomic number: the higher the atomic number, the higher the priority. In terms of priority, if there is a hydrogen attached to a stereogenic centre, it must always have the lowest priority. Groups which have an oxygen atom attached directly to the stereogenic carbon (for example, —OH, —OCH$_3$) have higher priority than groups which have a nitrogen atom attached directly to the stereogenic centre (for example, —NH$_2$, —N(CH$_3$)$_2$), and so on.

$$\text{Increasing order of priority: H < C < N < O < Cl < Br < I}$$

(ii) If two of the atoms attached to the stereogenic centre are identical (for example, —CH$_3$, —CH$_2$CH$_3$), then look to the next atoms away from the stereogenic centre. If the atoms at this point are still the same, then move further from the stereogenic centre (and so on) until the point of difference is found (there must be a point of difference somewhere for the molecule to be chiral). The group priority is then assigned based on the atomic numbers of the atoms at the first point of difference.

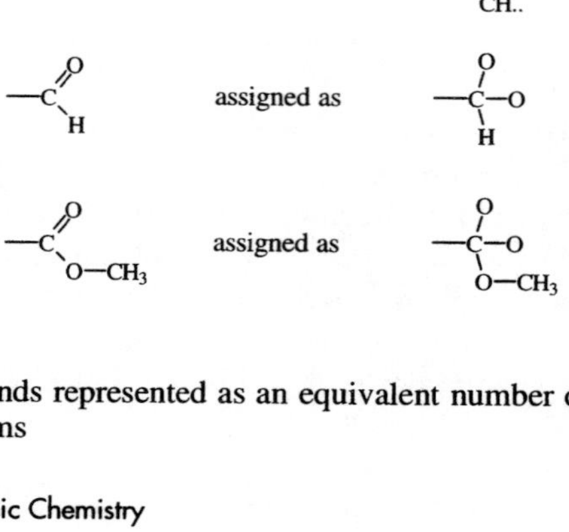

Figure 3.23 Multiple bonds represented as an equivalent number of single bonds to the same type of atoms

So, for example, —CH$_2$Cl has a higher priority than —CH$_2$OH because Cl has a higher priority than O. For example,

$$-CH_3 < -CH_2CH_3 < -CH(CH_3)_2 < -CH_2Cl < -CH_2Br$$

(iii) For the purposes of assigning group priorities, multiple bonds are expanded to be an equivalent number of single bonds attached to similar atoms. So a —CH=CH$_2$ group is considered as equivalent to —CH(—CH$_2$)$_2$ (Figure 3.23).

PRACTICE EXERCISE

In the following structure, arrange the groups attached to the stereogenic centre in increasing (lowest to highest) order of priority:

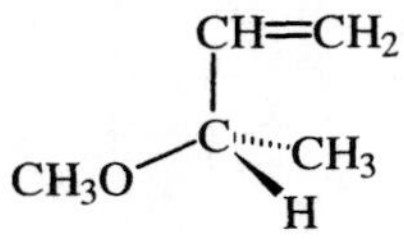

Answer: The —H substituent has the lowest priority. Of the other substituents attached to the stereogenic centre, two have carbon atoms as the point of attachment and one has an oxygen atom as the point of attachment. Oxygen has a higher atomic number than carbon so the —OCH$_3$ has the highest priority. The remaining two groups cannot be distinguished based on the priority of the atom directly attached to the stereogenic centre — carbon in both cases. The alkenyl group (—CH=CH$_2$) is now counted as —CH(—CH$_2$)$_2$ for the purpose of establishing group priorities. Look at the next atom away from the stereogenic centre: H for the methyl group and C for the alkenyl group. The C has priority over the H, so the groups arranged in increasing priority order are:

$$-H < -CH_3 < -CH=CH_2 < -OCH_3$$

SAMPLE EXERCISE 3.9

In the following structure, arrange the groups attached to the stereogenic centre in increasing (lowest to highest) order of priority:

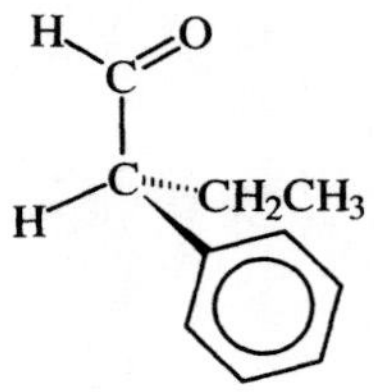

Answer:

—H < —CH$_2$CH$_3$ < —C$_6$H$_5$ < —C(H)=O

Once the priority sequence has been assigned, the configuration of the stereogenic centre is assigned by viewing the molecule from a position *opposite to* the group of lowest priority. Looking down the axis of the bond between carbon and the group of lowest priority, if the imaginary path traced out in moving from the group of highest priority to the group of second priority to the group of third priority traces a clockwise direction, the stereogenic centre is assigned the *R* configuration. If the path traced out in moving from the group of highest priority to the group of second priority to the group of third priority traces an anti-clockwise direction, the stereogenic centre is assigned the *S* configuration (Figure 3.24).

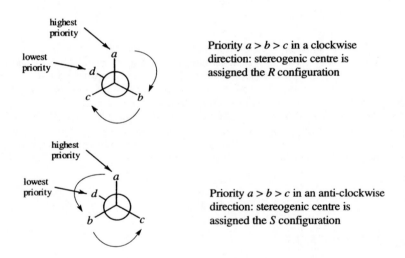

Figure 3.24 Assigning *R* and *S* configurations to a stereogenic centre

SAMPLE EXERCISE 3.10

Does the stereogenic centre in the compound whose structure appears below have the *R* or the *S* configuration?

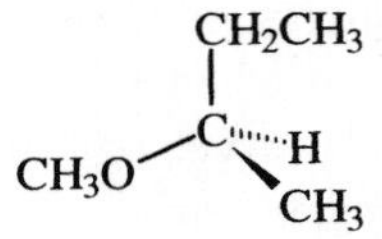

$$\underset{CH_3O}{}\overset{CH_2CH_3}{\underset{CH_3}{\overset{|}{C}\cdots H}}$$

Solution: The groups attached to the stereogenic centre can be assigned the following priority order:

$$-H < -CH_3 < -CH_2CH_3 < -OCH_3$$

If the molecule is arranged so that it can be viewed down the C—H bond with H (the lowest priority group) at the back of the molecule, away from the observer, then the stereogenic centre appears:

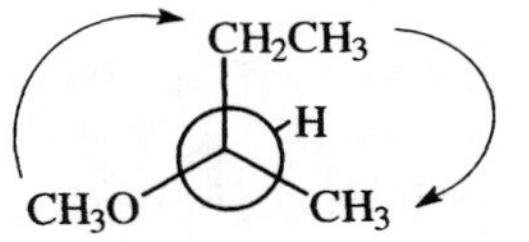

The path traced out in moving from the highest priority group to the group of second highest priority to the group of third priority traces a clockwise direction and this indicates that the stereogenic centre has the *R* configuration. The full name of this compound would be (*R*)-2-methoxybutane.

PRACTICE EXERCISE

Draw the enantiomer of 2–chloro-1–butanol which has the *S* configuration.

Answer:

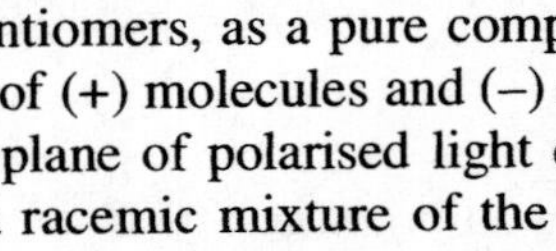

$$\underset{HO-CH_2}{}\overset{Cl}{\underset{CH_2CH_3}{\overset{|}{C}\cdots H}}$$

A mixture that contains equal amounts of two enantiomers of the same chiral compound is called a **racemic mixture**, and shows no rotation when placed in a polarimeter. Even though either of the enantiomers, as a pure compound, is optically active, the fact that there are equal numbers of (+) molecules and (–) molecules in a racemic mixture means that their effects on the plane of polarised light exactly cancel to give no net rotation. A compound which is a racemic mixture of the two enantiomers is given the symbol (±) when referring to its optical rotation or the prefix *R,S*- when giving it a formal name.

In a chemical reaction which begins with achiral starting materials (that is, molecules containing no stereogenic centre) and employs achiral reagents it is not possible to generate a product which is a pure *R*- or *S*- enantiomer. If the reaction begins with achiral starting materials and generates a compound which contains a stereogenic centre, then equal numbers of molecules with *R*- and *S*- configurations at that centre will be formed and the product must be a racemic mixture.

For example, when butanone is treated with LiAlH$_4$ followed by acidification, 2–butanol is produced. The starting materials do not contain any stereogenic centres but the product has a stereogenic centre.

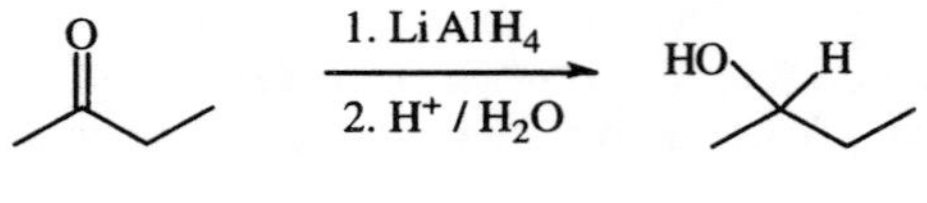

The carbon atom of the carbonyl group of butanone is *sp*^{2} hybridised and there is a trigonal planar arrangement of the groups attached to it. The hydride nucleophile will attack the carbon atom in a nucleophilic addition reaction to form a new bond between the hydride and the carbon and break the π bond. Since the carbonyl carbon is planar, there is absolutely no difference between the upper and lower faces of the plane of the carbonyl group and attack from either face is equally likely. Thus, when the new stereogenic centre is generated, equal numbers of molecules will be generated with the *R*- and the *S*- configurations and the result will be a product which is a racemic mixture.

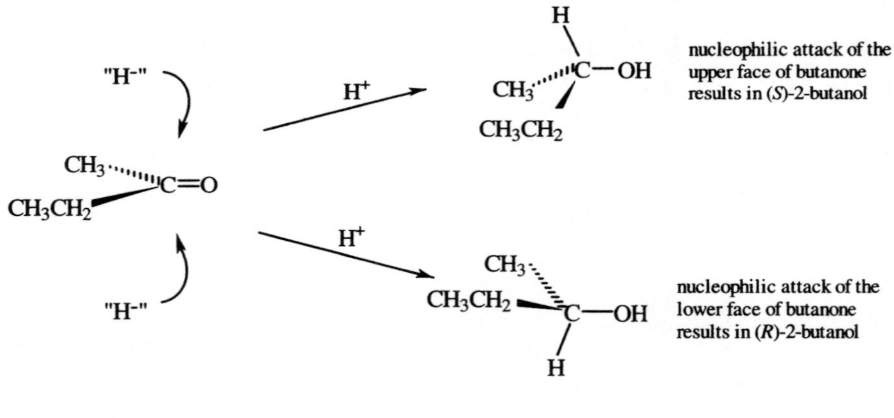

nucleophilic attack of the upper face of butanone results in (*S*)-2-butanol

nucleophilic attack of the lower face of butanone results in (*R*)-2-butanol

If the starting material contains a chiral carbon atom or if the reagent which performs the reaction is chiral then it is possible to produce products which are pure enantiomers or at least enriched in one enantiomer. In living systems, the compounds that make up cells and cellular fluids are typically chiral and the reagents which make biological molecules or transform them into other biological molecules are also chiral. In most natural systems, the molecules are synthesised as single enantiomers.

Molecules with more than one stereogenic centre

Each stereogenic centre in a molecule can be assigned the absolute configuration R or S depending on the arrangement of the groups or atoms around it. For compounds with only one stereogenic centre, the compound with an R configuration is the enantiomer of the compound with the S configuration. For compounds with more than one stereogenic centre, the situation is more complex. There are four possible isomers of a compound with two stereogenic centres. These arise from the fact that the absolute configuration at one of the centres can be R or S and for the compound with the R configuration, there are isomers with either the R or the S configuration at the other centre. Similarly for the compound with the S configuration at the first centre, there are isomers with either the R or the S configuration at the other centre (Figure 3.25).

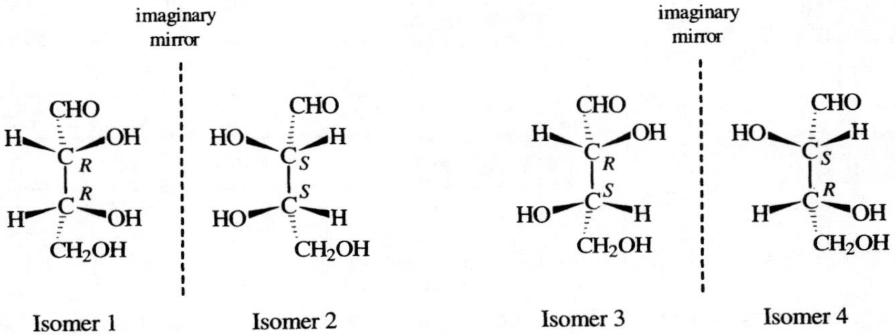

Figure 3.25 There are four stereoisomers of a molecule with two stereogenic centres

The relationship between pairs of compounds with two stereogenic centres is illustrated in Figure 3.25. The compound with the R configuration at C2 and the R configuration at C3 (isomer 1) is the mirror image of the compound with the S configuration at C2 and the S configuration at C3 (isomer 2). Isomers 1 and 2 are a

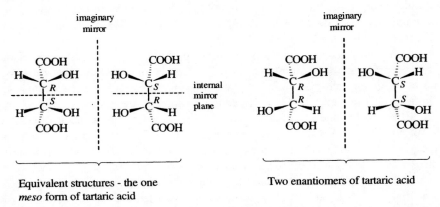

Equivalent structures - the one *meso* form of tartaric acid

Two enantiomers of tartaric acid

Figure 3.26 The three stereoisomeric forms of tartaric acid—two enantiomers and a *meso* isomer

Separation and Identification of Organic Compounds

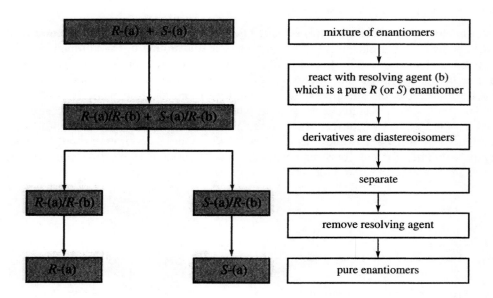

Figure 3.27 Scheme for resolving enantiomers by forming diastereoisomers

pair of enantiomers and similarly isomers 3 and 4 are a pair of enantiomers. However isomers 1 and 3 (or the pairs of isomers 1 and 4, 2 and 3, or 2 and 4) are not enantiomers, but diastereoisomers. Recall that diastereoisomers are stereoisomers which are not mirror images of each other. The distinction between compounds which are enantiomers and those which are diastereoisomers is important. Enantiomers have identical physical and chemical properties, with the exception of the direction of rotation of plane polarised light, whereas diastereoisomers do not have identical physical and chemical properties. A pair of diastereoisomers can be separated by the normal methods (for example, recrystallisation, chromatography) whereas enantiomers cannot. So for the compounds listed in Figure 3.25, isomer 1 could be readily separated from isomers 3 and 4, but not from isomer 2.

If there are n stereogenic centres in a molecule, there are 2^n possible stereoisomers. The only exception to this general rule arises when a molecule which possesses more than one stereogenic centre contains a plane of symmetry and *is* superposable on its mirror image. This occurs when one stereogenic centre is the mirror image of another in the same molecule. These compounds are called *meso compounds*. In this case the number of possible stereoisomers is reduced from 2^n. Tartaric acid (2,3–dihydroxy-butanedicarboxylic acid) has two stereogenic centres (at C2 and C3). The two stereogenic centres have identical substituents and one of the possible stereoisomers of this compound has an internal mirror plane such that the arrangement of groups at C2 is the mirror image of those at C3. This stereoisomer is called the **meso isomer** of tartaric acid (Figure 3.26). Despite the fact that a *meso* compound has stereogenic centres within its structure, they appear optically inactive. This is because the rotation of plane polarised light by one stereogenic centre within a molecule is equal and opposite to the rotation caused by the other mirror image centre, within the same molecule, and the two effects exactly cancel.

Resolution of a mixture of enantiomers

The separation of a mixture of enantiomers from one another is called **resolution**. The usual method for resolving a mixture into its components is to chemically introduce an additional stereogenic centre into the molecule to convert the molecules that are enantiomers into derivatives that are diastereoisomers. The diastereoisomers can be separated and then the resolving agent removed to give the pure enantiomers (Figure 3.27).

Resolving agents are pure enantiomers typically derived from natural sources. In the process of resolution, the resolving agent is chemically attached to permit the separation and it is then removed. It can usually be recovered (and reused). Common resolving agents include naturally occurring carboxylic acids, amines, and alcohols which can be chemically attached to a molecule by the formation of a stable carboxylic acid derivative (for example, ester, amide) to effect the separation then removed simply by hydrolysis.

SUMMARY

Each year many thousands of new or previously unknown organic compounds are discovered and have to be identified. Before they can be identified they must be purified or separated from other compounds. Separation of liquids can be achieved by distillation, using a fractionating column if necessary. Recrystallisation can be used to separate solids or if the solid is relatively volatile it can be purified by sublimation. Compounds with acidic or basic functional groups, such as amines and carboxylic acids, can be separated by acid/base extraction which uses the differential solubility of charged and uncharged species in water and non-aqueous solvents. Chromatography is a general method for separating organic compounds and it makes use of the different affinities compounds have for solids, such as alumina, silica, or natural or synthetic polymers. Gas chromatography (GC) and high pressure liquid chromatography (HPLC) are widely used in fields such as quality control and environmental analysis.

The empirical formula of a molecule is determined by elemental analysis and, when coupled with a molecular weight measured by a mass spectrometer, the molecular formula can be calculated. Identification of the functional groups in an organic compound is most frequently achieved by making use of one or more types of spectroscopy. A mass spectrometer separates compounds on the basis of their molecular weights and the breakdown of a molecule into characteristic fragments in a mass spectrometer can also be used to aid the identification. UV-visible absorption spectroscopy is limited to compounds with conjugated double or triple bonds. IR spectra have absorptions that arise from bond vibrations and the bending of bond angles. The frequency of these absorptions can be used to determine the presence of certain functional groups. The fingerprint region of an IR spectrum is different for every compound and can be used for identification by comparison of the spectrum of the unknown compound with a library of these fingerprints. Nuclear magnetic resonance

(NMR) spectroscopy detects various elements in a compound (commonly hydrogen) on the basis of the absorption of Rf radiation by their nuclei. For ^{1}H NMR spectroscopy, the frequency of the absorption is highly sensitive to the chemical environment of the hydrogen atom and so the spectrum can help reveal what types of functional groups are nearby. Also, the splitting of a ^{1}H NMR signal arising from an atom is dependent on the number and type of surrounding atoms so in many cases the NMR spectrum can be used to determine the structure of the compound. A diffraction analysis of crystals can acurately determine bond distances and angles.

Isomerism is common in organic compounds so a knowledge of the chemical formula alone is often not sufficient to fully identify a compound. Constitutional isomers have the same chemical formula, but a different atom connectivity. Stereoisomers have the same chemical formula and connectivity, but the arrangements of the bonds and atoms in space differs. Stereoisomers which are non-superposable mirror images of each other are called *enantiomers*, those that are not mirror images of each other are called *diastereoisomers*. Enantiomers are also referred to as *optical isomers*. Enantiomers usually contain an sp^3 hybridised atom with four different groups bonded to it. This atom is referred to as a stereogenic centre or chiral carbon centre. A pair of enantiomers differ from one another in the direction in which they rotate plane-polarised light and this can be identified in a polarimeter. The absolute configuration of a chiral carbon centre refers to the exact three-dimensional arrangement of groups around it and is labeled in the compound name on the basis of the priorities of the groups.

KEY TERMS

absolute configuration
absorption spectroscopy
acid/base extraction
chemical shift
chiral carbon centre
chiral centre
chirality
chromatography
column chromatography
conjugated
crystallisation
daughter ion
diastereoisomer
distillation
enantiomer
fingerprint region
fractionating column
gas chromatography
high pressure liquid chromatography
identification

Infrared spectroscopy
isotopic abundances
mass spectrometry
meso compounds
meso isomer
multiplicity
nuclear magnetic resonance spectroscopy
optically active
parent ion
Polarimeter
racemic mixture
recrystallisation
resolution
separation
sequence rules
specific rotation
stereogenic centre
stereoisomer
sublimation
UV-visible spectroscopy

EXERCISES

Separation and purification of organic compounds

1. Describe a method you could use to separate and purify 1–octylamine from a mixture containing 1–octylamine, 1–chlorooctane, and ethyl acetate.
2. Describe a method you could use to separate and purify benzoic acid from a mixture containing benzoic acid, *N,N*-dimethylbenzamide, and decanol.

Mass spectrometry

3. Which of the following compounds would have molecular ions in their mass spectrum with *m/z* 86?

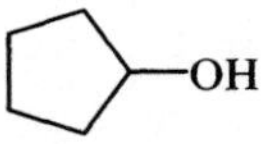

(a) cyclopentanol (–OH on cyclopentane)

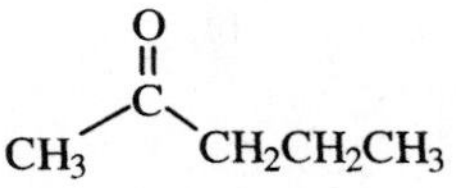

(b) $CH_3-C(=O)-CH_2CH_2CH_3$

(c) $CH_3-CH_2-CH_2-CH_2-C(=O)-H$

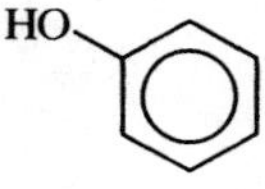

(d) HO–phenyl (phenol)

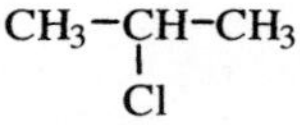

(e) $CH_3-CH(Cl)-CH_3$

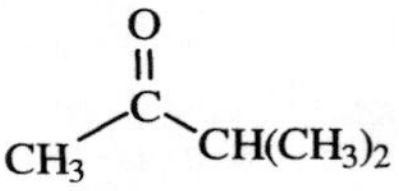

(f) $CH_3-C(=O)-CH(CH_3)_2$

4. Which of the following compounds would give fragment ions in their mass spectrum with *m/z* 43 ?

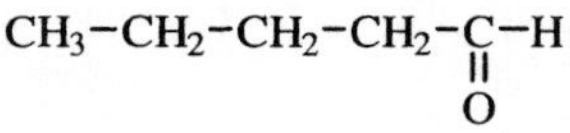

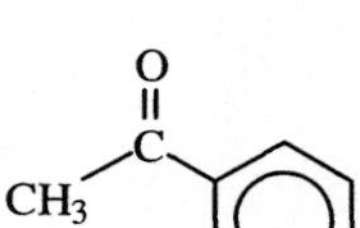

(a) $CH_3-C(=O)-$phenyl

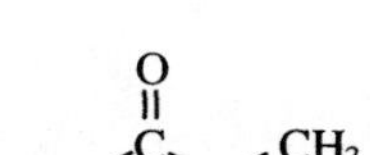

(b) $CH_3-C(=O)-O-CH_3$

(c) $CH_3-CH_2-CH_2-CH_3$

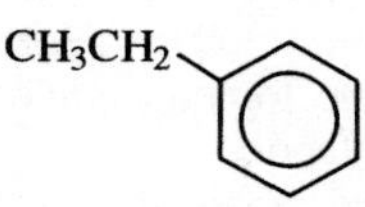

(d) CH_3CH_2–phenyl

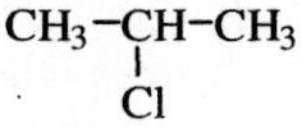

(e) $CH_3-CH(Cl)-CH_3$

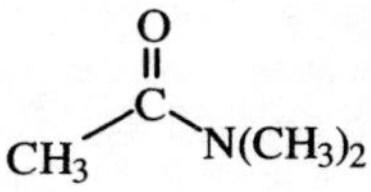

(f) $CH_3-C(=O)-N(CH_3)_2$

5. Which of the following compounds would give fragment ions in their mass spectrum with *m/z* 91?

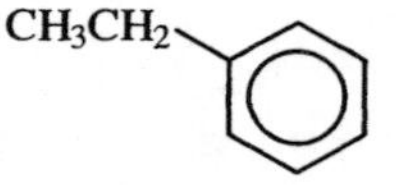

(a) CH₃CH₂-C₆H₅

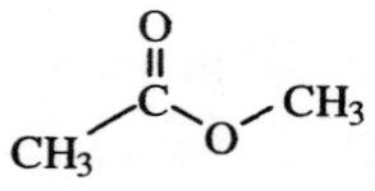

(b) CH₃-C(=O)-O-CH₃

(c) CH₃-CH₂-Br

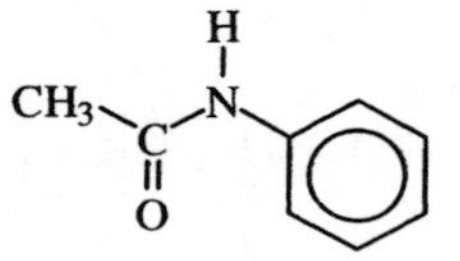

(d) CH₃-C(=O)-N(H)-C₆H₅

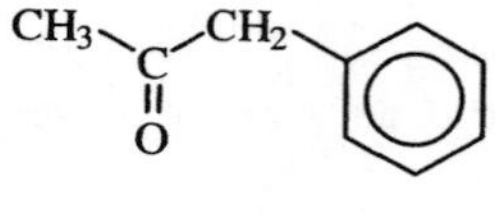

(e) CH₃-C(=O)-CH₂-C₆H₅

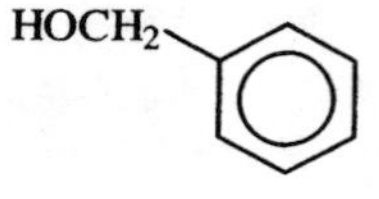

(f) HOCH₂-C₆H₅

6. Which of the following compounds would have ions in their mass spectrum at both m/z 122 and m/z 124 of almost equal intensity? Why?

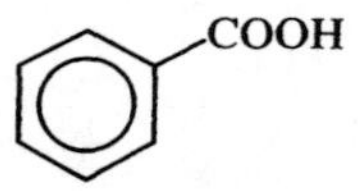

(a) C₆H₅-COOH

(b) CH₃-CH₂-CH₂-Br

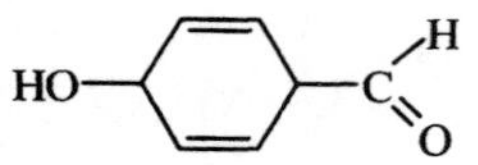

(c) HO-C₆H₄-CHO

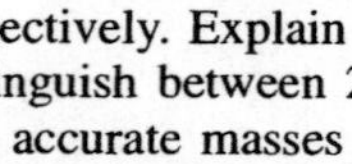

(d) CH₃-CH(Br)-CH₃

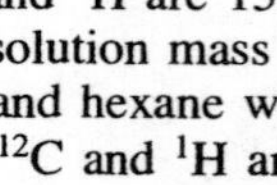

(e) cyclopentyl-Br

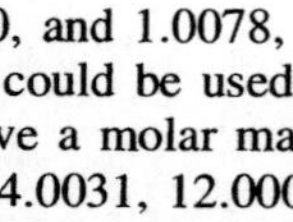

(f) cyclohexyl-Cl

7. The accurate masses of ¹⁶O, ¹²C and ¹H are 15.9949, 12.000, and 1.0078, respectively. Explain how high resolution mass spectometry could be used to distinguish between 2–pentanone and hexane which both have a molar mass of 86.

8. The accurate masses of ¹⁶O, ¹⁴N, ¹²C and ¹H are 15.9949, 14.0031, 12.000, and 1.0078 respectively. Explain how high resolution mass spectometry could be used to distinguish between 4–hydroxy-2–butanone and 3–pentanol.

9. The mass spectrum below arises from a compound whose molecular formula is C₅H₁₀O. Give the structural formula of the molecule from which the spectrum was obtained and give reasons for your answer.

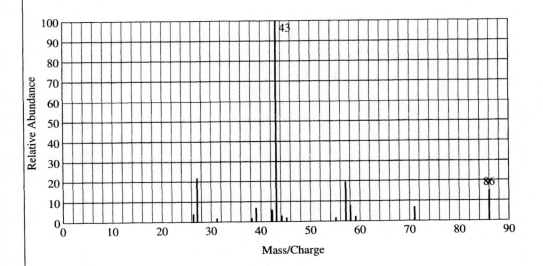

UV-visible spectroscopy

10. Which of the following compounds would absorb radiation strongly in the UV-visible range of the electromagnetic spectrum?

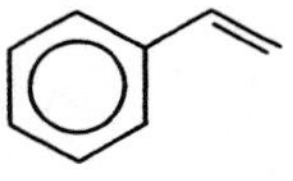

(a) styrene

(b) CH₃—CH₂—CH₂—Br

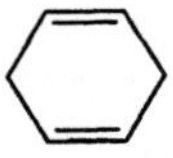

(c) cyclohexadiene

(d) CH₃—CH₂—CH₃

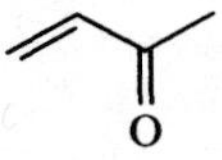

(e) methyl vinyl ketone

(f) hex-1-en-3-yn-5-ene

11. A compound with molecular formula C_5H_8O has a strong absorption in the UV spectrum at 270 nm. Give the condensed structural formula of a compound which would fit this data.

Infrared spectroscopy

12. Which of the following compounds would absorb radiation strongly in the ranges: (i) 3200 to 3600 cm^{-1}? (ii) 1680 to 1750 cm^{-1}?

Separation and Identification of Organic Compounds **165**

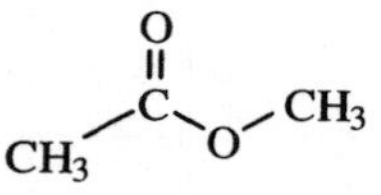

(a) CH₃−C(=O)−O−CH₃

(b) CH₃−CH₂−Br

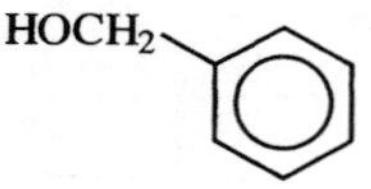

(c) HOCH₂−C₆H₅

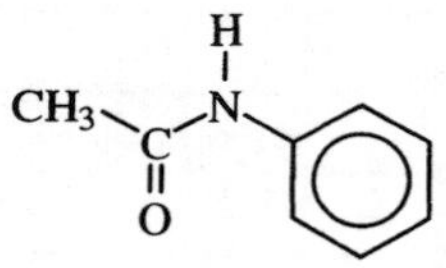

(d) CH₃−C(=O)−N(H)−C₆H₅

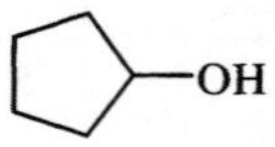

(e) cyclopentyl−OH

(f) CH₃−CH₂−NH−CH₃

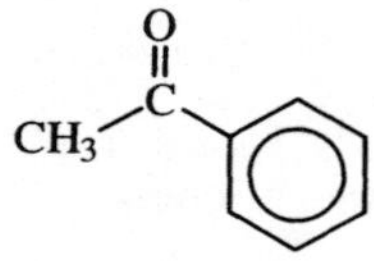

(g) CH₃−C(=O)−C₆H₅

(h) CH₃CH₂CH₂−C(=O)−H

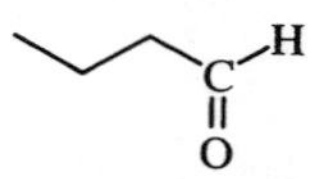

13. The infrared spectrum below arises from a compound whose molecular formula is $C_5H_{12}O$. Give possible structural formulas of the molecules from which the spectrum was obtained and give reasons for your answer.

14. A compound with the molecular formula C_3H_8O has *no* absorption in the IR spectrum in the region 3200 to 3600 cm^{-1} and *no* strong absorption in the UV region of the electromagnetic spectrum. Give the structural formula of the compound and give reasons for your answer.

15. A compound with the molecular formula C_5H_8O has a strong absorption in the IR spectrum at 1720 cm^{-1} and *no* strong absorption in the UV region of the electromagnetic spectrum. Give possible structural formulas of compounds which would be consistent with this data and give reasons for your answer.

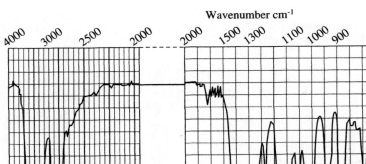

Infrared spectrum

NMR spectroscopy

16. Which of the following compounds would have a ^{1}H NMR spectrum consisting of: (i) only one singlet signal? (ii) only two singlet signals? (iii) only a triplet and a quartet signal?

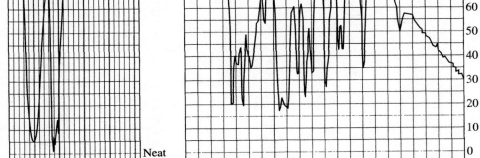

17. For the ¹H NMR spectrum of each of the following compounds, give the number of signals you would expect to see, the relative intensities of the signals, and the multiplicities of each of the signals:

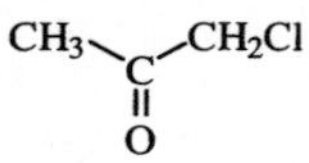

(a) CH₃−C(=O)−CH₂Cl

(b) CH₃−CH₂−Cl

(c) CH₃CH₂−C(=O)−O−CH₃

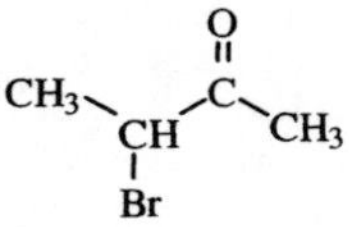

(d) CH₃−CH(Br)−C(=O)−CH₃

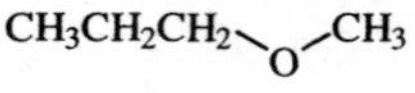

(e) CH₃CH₂CH₂−O−CH₃

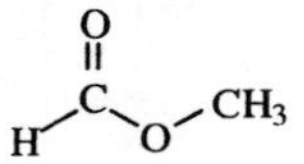

(f) H−C(=O)−O−CH₃

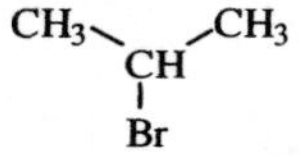

(g) CH₃−CH(Br)−CH₃

18. For the ¹H NMR spectra of each of the following compounds, give the number of signals you would expect to see, the relative intensities of the signals, and the multiplicities of each of the signals:

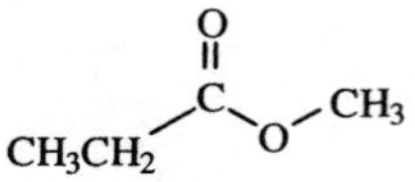

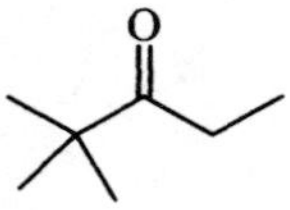

(a) (CH₃)₃C−C(=O)−CH₂CH₃

(b) BrCH₂−C(=O)−O−CH₃

(c) CH₃CH₂−C(=O)−O−CH₂CH₃

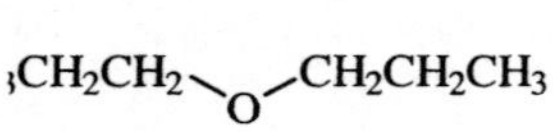

(d) CH₃CH₂−O−CH₂CH₂CH₃

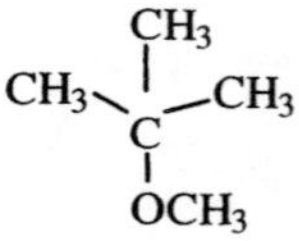

(e) CH₃−C(CH₃)(OCH₃)−CH₃

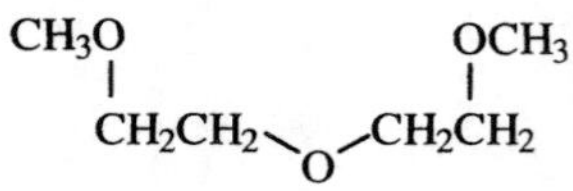

(f) CH₃O−CH₂CH₂−O−CH₂CH₂−OCH₃

(g) Cl−CH₂−CH₂−Cl

168 The Essentials of Organic Chemistry

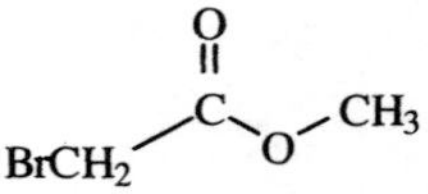

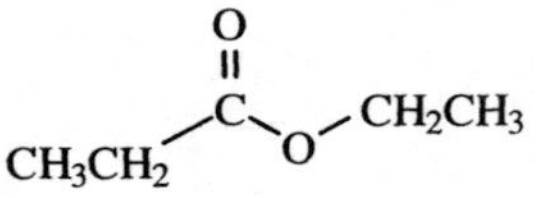

19. A compound with the molecular formula C_4H_8O has a strong absorption in the IR spectrum at 1720 cm^{-1} and a ^1H NMR spectrum consisting of a triplet, a singlet, and a quartet signal. Give the structural formula of a compound which would be consistent with this data and give reasons for your answer.
20. A compound with the molecular formula $C_5H_{10}O_2$ has a strong absorption in the IR spectrum at 1750 cm^{-1} and a ^1H NMR spectrum consisting of two quartets and two triplets. Give the structural formula of a compound which would be consistent with this data and give reasons for your answer.

Optical isomerism

21. Identify the stereogenic centre in the following compounds:

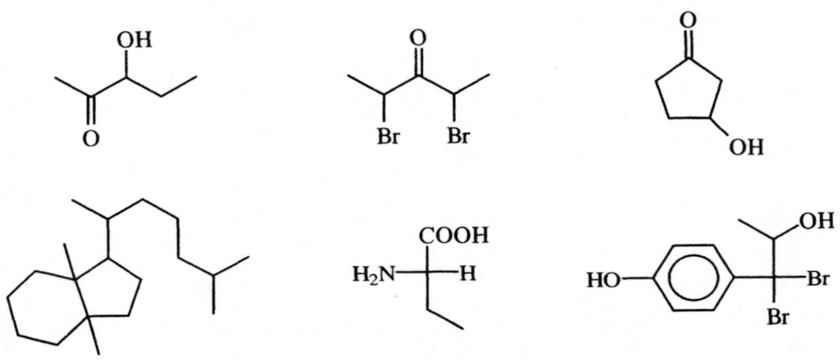

22. Draw the enantiomer of each of the following compounds:

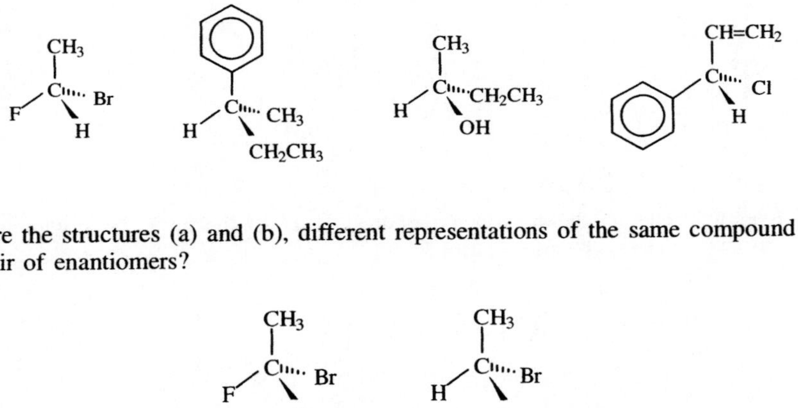

23. Are the structures (a) and (b), different representations of the same compound or a pair of enantiomers?

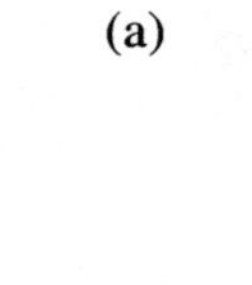

(a) 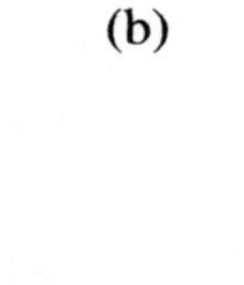 (b)

Separation and Identification of Organic Compounds **169**

24. Are the structures (a) and (b), different representations of the same compound or a pair of enantiomers?

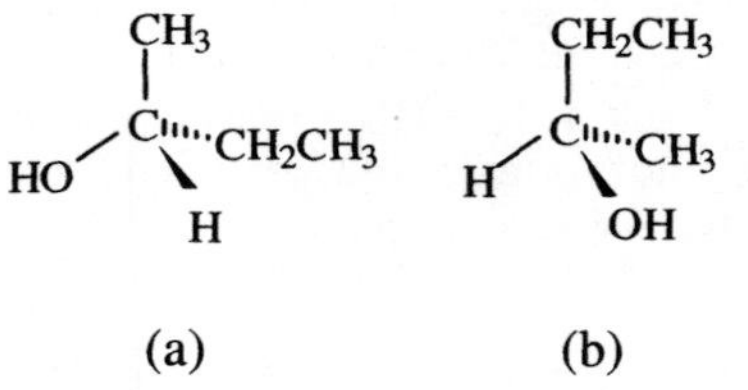

(a)   HO—C(CH₃)(CH₂CH₃)(H) [wedge structure]

(b)   H—C(CH₂CH₃)(CH₃)(OH) [wedge structure]

25. In each of the following molecules, label the groups attached to the stereogenic centre in order of *decreasing* (highest to lowest) priority:

(a) HO—C(CH₃)(H)(F)

(b) HO—C(CH₃)(CH₂CH₃)(H)

(c) Ph—C(CH=CH₂)(H)(CH₃)

(d) CH₃O—C(CH₃)(CF₃)(OCH₂CH₃)

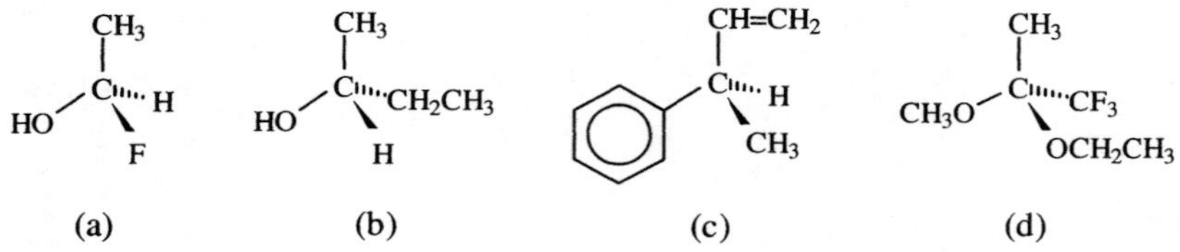

26. In each of the following molecules, label the groups attached to the stereogenic centre in order of *decreasing* (highest to lowest) priority:

(a) OHC—C(CH₃)(Br)(F)

(b) HO—C(CH₃)(CH₂CH₃)(CH₂CH₂CH₃)

(c) Ph—C(CH=CH₂)(C(=O)CH₃)(CH₃)

(d) CH₃O—C(CH₃)(CH₂CH₃)(C(CH₃)₂H)

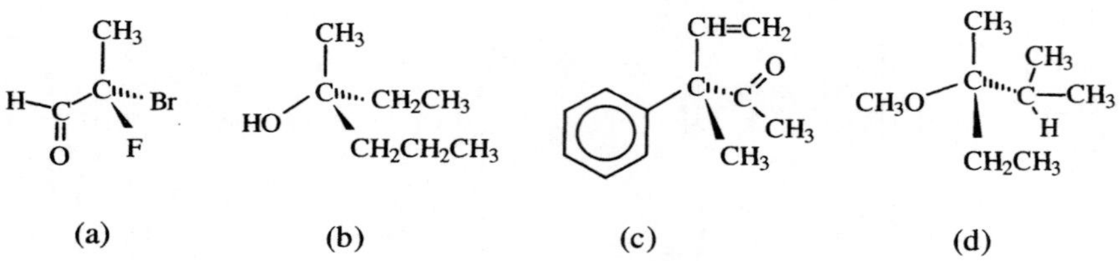

27. Establish whether the stereogenic centre in each of the following compounds has the *R*— or the *S*— configuration:

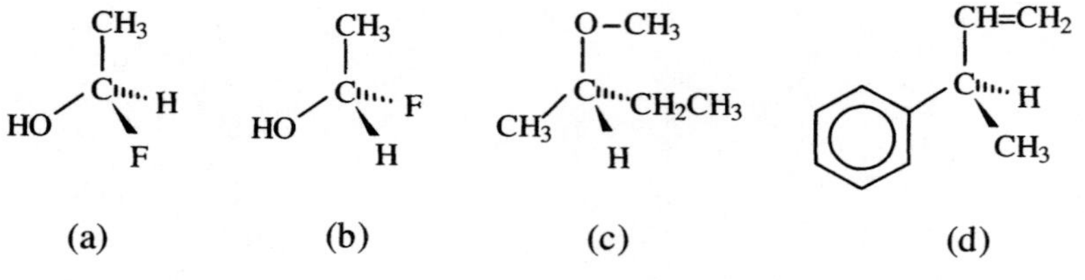

(a) HO—C(CH₃)(H)(F)

(b) HO—C(CH₃)(F)(H)

(c) CH₃—C(O—CH₃)(CH₂CH₃)(H)

(d) Ph—C(CH=CH₂)(H)(CH₃)

170 The Essentials of Organic Chemistry

28. Give the structure of: (i) (*R*)-2-methyl-3-pentanol; (ii) (*S*)-1-phenylethylamine; (iii) (2*R*, 3*S*)-2,3-dihydroxybutane.
29. Which of the following are *meso* compounds?

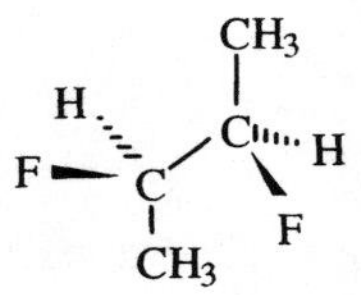

(a) 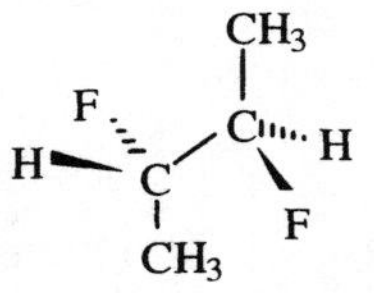 (b) 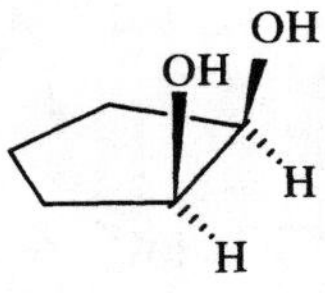 (c) 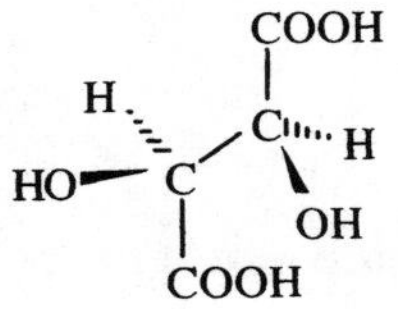 (d)

30. The compound *(S)*-2-hydroxypropionic acid (also called *(S)*-lactic acid) has a specific rotation of [α] = +2.3°. What is the specific rotation of *(R)*-2-hydroxypropionic acid?
31. Which of the following compounds shows optical activity?
    (i) *(R)*-2-methyl-3-pentanol; (ii) 3,3-dimethyl-1-butanol; (iii) (2*R*,3*S*)-2,3-dihydroxy-1,4-butanedicarboxylic acid; (iv) *meso*-1,2-dihydroxylcyclopentane; (v) (±)-2-butanol.
32. Reaction of 1–propene with Br_2 in carbon tetrachloride solvent produces 1,2–dibromopropane. Is the compound obtained as an *R*-enantiomer, an *S*-enantiomer, an achiral compound, a racemic mixture or a *meso* compound?
33. Reaction of 2–pentanone with $LiAlH_4$ in dry ether solvent followed by acidification with aqueous acid, produces 2–pentanol. Is the compound obtained as an *R*-enantiomer, an *S*-enantiomer, an achiral compound, a racemic mixture or a *meso* compound?
34. Hydrolysis of the ester (a) with dilute acid produces a carboxylic acid and an alcohol. Is the carboxylic acid obtained as an *R*-enantiomer, an *S*-enantiomer, an achiral compound, a racemic mixture or a *meso* compound?

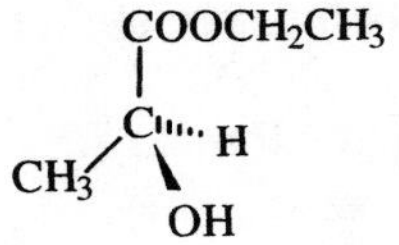

(a)

Glossary

Absolute configuration	The three-dimensional arrangement of the groups about a chiral centre.
Absorption spectroscopy	Spectroscopy arising from the absorption of electromagnetic radiation.
Acetal	A functional group with two ether oxygen atoms singly bonded to the same carbon atom, $RCH(OR')_2$.
Acid Anhydride	A carboxylic acid derivative with two acyl groups linked by a common oxygen, $(RCO)_2O$.
Acid/base extraction	Separation of weakly acidic or weakly basic compounds by changing their solubility as a result of varying pH.
Acid Chloride	Carboxylic acid derivative with a chlorine attached to an acyl group, $RCOCl$.
Addition reaction	The addition of one molecule to another to form one product.
Alkane	A molecule containing only carbon and hydrogen bonded with single σ-bonds.
Alcohol	A molecule containing at least one alkyl carbon atom bearing a hydroxy group, ROH.
Alkene	An hydrocarbon molecule containing at least one carbon-carbon double bond.
Aldehyde	A carbonyl group bonded to a hydrogen and a carbon group, $RCHO$.
Alkoxide ion	The anion formed (RO^-) by removing H^+ from an alcohol, ROH.
Alkyl Halide	A molecule containing at least one alkyl carbon atom bearing a halogen atom, RX.
Alkyl group	A hydrocarbon fragment containing only carbon and hydrogen bonded with single σ-bonds.
Alkyne	A hydrocarbon molecule containing at least one carbon-carbon triple bond.
Amide	A carboxylic acid derivative containing a nitrogen atom singly bonded to an acyl group, $RCONH_2$, $RCONHR'$, $RCONR'_2$.
Amine	A molecule containing a nitrogen atom singly bonded to carbon or hydrogen, RNH_2, $RNHR$, RNR_2.
Ammonium ion	The cation formed when nitrogen is bonded to four carbon or hydrogen atoms, NH_4^+, NR_4^+, NHR_3^+ *etc*.

Aromatic diazonium salts	A salt which contains dinitrogen bonded to an aromatic ring, $A_rN_2{}^+X^-$.
Anti	A label used to describe a conformation of a hydrocarbon.
Aromatic Hydrocarbon	A stable cyclic hydrocarbon molecule containing conjugated double bonds, *e.g.* benzene. The π-electrons in an aromatic hydrocarbon are delocalised giving the molecule unusual stability.
Aromatic electrophilic substitution	The dominant reaction of simple aromatic hydrocarbons where a hydrogen on an aromatic ring is replaced by an electrophile.
Aryl Halide	A molecule in which a halogen is bonded to an aromatic ring.
Carbocation	A cation in which a carbon atom bears a positive charge.
Carbonyl group	A functional group comprising of carbon doubly bonded to an oxygen.
Carboxylic Acid	A functional group comprising carbon doubly bonded to one oxygen and singly bonded to an $-OH$.
Chemical shift	The position (or frequency) of a signal in an NMR spectrum.
Chirality	The condition that a compound and its mirror image are not superposable.
Chiral carbon centre	The carbon centre that gives rise to chirality. In organic compounds the most common chiral centres are carbon atoms with four different groups attached. A chiral centre is also termed a stereogenic centre.
Chromatography	A technique for separating compounds based on their tendency to be adsorbed on a solid support.
Chromatography column	A column or tube packed with a solid adsorbent. Organic compounds are differentially absorbed when passed through a chromatography column in the gas phase or in solution.
Cis/Trans Isomers	Isomers which differ in the relative positions of groups attached to a double bond or a ring system.
Combustion	Complete reaction of an organic compound with oxygen to form carbon dioxide and water.
Configuration	The arrangement of groups bonded to a carbon atom.
Conformational Isomers	Molecules whose structures differ only by rotation about single bonds.
Conjugated	An alternating arrangement of single bonds and double or triple bonds.
Constitutional Isomers	Molecules with the same molecular formula but a different sequence of bonds or atoms.
Crystallisation	Formation of a highly ordered solid from a saturated solution.
Cycloalkane	A cyclic hydrocarbon.

Daughter ion	Peaks in a mass spectrum arising from fragmentation of molecules in a mass spectrometer.
Diastereoisomers (Diastereomers)	Stereoisomers which are not mirror images of one another.
Diazonium Ion	A cation containing an N_2^+ group.
Diol	A molecule containing two hydroxyl groups.
Directing Influence	The ability of a group already attached to an aromatic ring to favour substitution at specific positions of that ring.
Distillation	The separation or purification of the volatile components of a mixture by boiling off of each component in turn.
Eclipsed	A label used to describe a conformation of a hydrocarbon, usually a high energy form.
Electrophile	A species which reacts with an electron-rich centre.
Elimination reaction	Splitting off a small molecule or group to form an unsaturated product.
Enantioners	Stereoisomers that are mirror images of each other, but not superposable.
Ester	A carboxylic acid derivative in which the —OH is replaced by an alkoxy group (—OR), R'COOR.
Ether	A molecule containing oxygen singly bonded to two carbon atoms, ROR'.
Fingerprint region	The region of an infrared spectrum below approximately $1500\ cm^{-1}$. The fingerprint region is important because this region is unique for each different compound and can be used as a means of identification or 'fingerprint'.
Fractionating column	A column usually packed with glass beads, used in the separation of volatile compounds by distillation. A fractionating column improves the separation of components with similar boiling points.
Free Radical	A species possessing an unpaired electron.
Friedel Crafts acylation	The introduction of an acyl group (RC=O) into an aromatic hydrocarbon via an electrophilic substitution reaction.
Friedel Crafts alkylation	The introduction of an alkyl group into an aromatic hydrocarbon.
Functional group	A group of atoms which confer characteristic chemical and physical properties to the compounds which contain them.
Gas chromatography	Chromatography in which the compounds are passed through a chromatography column in the gas phase.
Gauche	A label used to describe a conformation of an alkane.
Grignard Reagent	An alkyl- or aryl-magnesium halide formed by the reaction of metallic magnesium with an alkyl or aryl halide.
Hemiacetal	An alcohol and an ether group attached to the same carbon atom, RCH(OR')(OH).
Heteroatom	An atom other than carbon or hydrogen, *e.g.* N, O, F, Cl, Br, I, S, Si, P.

High pressure liquid chromatography (HPLC)	Chromatography in which the compounds are passed through a chromatography column in solution and under high pressure.
Homologous Series	A family of molecules with related structures differing only by the number of —CH_2— units.
Hydrocarbon	A molecule containing only hydrogen and carbon atoms.
Hydrogen bond	A weak bond between an acidic hydrogen atom and the lone pair of a nucleophilic atom.
Imine	A functional group consisting of a nitrogen atom doubly bonded to a carbon and singly bonded to C or H.
Infrared spectroscopy	Absorption spectroscopy in which the radiation absorbed is in the infrared region of the electromagnetic spectrum.
Isotopic abundance	The proportion of the isotopes of each element present in a typical sample.
Ketone	A functional group consisting of a carbonyl group attached to two carbon atoms, RCOR'.
Markonikov's Rule	In the addition of HX to an alkene, the hydrogen adds to the alkene carbon with the greatest number of hydrogen atoms already attached.
Mass spectrometry	A technique for measuring the mass of molecules and fragments of the molecules. The method relies on ionisation of a compound in the gas phase then separating the charged products using a magnetic field.
***Meso* compound**	A stereoisomer with two chiral centres in which these centres are mirror images of each other.
Meta	Label signifying the relative position of two groups on a benzene ring. *Meta* indicates two groups are attached to carbon atoms separated by a single carbon atom.
Monomer	The molecule which forms the repeating units of a polymer.
Multiplicity	The number of peaks into which a signal in an NMR spectrum is split.
Nitration	Introduction of a nitro group (NO_2) into a molecule.
Nitrile	A functional group consisting of a nitrogen atom triply bonded to a carbon atom, RCN.
Nuclear magnetic resonance (NMR) spectroscopy	Spectroscopy based on the absorption of radio waves by the nuclei of atoms (most commonly hydrogen atoms) when a compound is placed in a strong magnetic field.
Nucleophilic substitution	A reaction in which a nucleophile displaces another atom or group in a molecule.
Nucleophile	A species which reacts with an electron-deficient centre.
Optically active	Molecules which rotate plane-polarised light are called *optically active* molecules. Entantiomers rotate the plane of polarised light in opposite directions but to the same extent.
Ortho	Label signifying the relative position of two groups on a

Ozonolysis	benzene ring. *Ortho* indicates two groups are attached to adjacent carbon atoms of a benzene ring. Cleavage of the C=C of an alkene with ozone (O_3).
Para	Label signifying the relative position of two groups on a benzene ring. *Para* indicates two groups are attached to opposite carbon atoms of a benzene ring.
Parent ion	The peak in a mass spectrum corresponding to the mass of the compound injected into the mass spectrometer.
Phenol	A compound containing a hydroxyl group attached to a benzene ring.
Phenoxide	The anion formed by removing H^+ from a phenol.
Polarimeter	An instrument for measuring the degree to which an optically active compound rotates the plane of polarised light.
Polymer	Large molecule composed of repeating groups of atoms.
Racemic mixture	A mixture containing both enantioners of a chiral compound in equal proportions.
Radical chain reaction	A self-perpetuation process which involves reaction of a radical with a substrate which in turn generates another radical.
Reaction mechanism	The sequence of steps by which one molecule is transformed into another.
Recrystallisation	Purification of a compound by slow induction of precipitation, leaving impurities behind in solution.
Resolution	Separation of the two enantioners in a racemic mixture.
Sequence rules	The rules used to assign priorities to different groups for the purpose of naming isomers or enantiomers. Sequence rules are also termed 'priority rules'.
Separation	The isolation of one component of a mixture of compounds from all of the other components.
Specific rotation	The rotation, in degrees, induced by a solution of a chiral compound (at a concentration of 1 g mL^{-1} in a 10 cm cell. The specific rotation (measured using a polarimeter) is a characteristic property of each chiral compound.
Staggered	A label used to describe a conformation of a hydrocarbon, usually with a low energy form.
Stereogenic centre	A carbon atom with four different groups attached to it.
Stereoisomer	Molecules with the same molecular formula and sequence of bonds but differing in the arrangement of the atoms or bonds in space.
Substitution reaction	A reaction in which one atom or group is replaced by another.
Sublimation	Transformation of a solid to a gas. Sublimation also refers to the purification of a volatile solid compound by heating (usually under reduced pressure) to vaporise it then condensation and collection of the solid.

UV-visible spectroscopy — Absorption spectroscopy in which the radiation absorbed is in the ultraviolet or visible region of the electromagnetic spectrum.

Z/E isomers — Isomers which differ in the relative positions of groups attached to a double bond.

Index

Absolute configuration 154
Absorption spectroscopy 135–147
α, β-unsaturated aldehydes 96
α, β-unsaturated ketones 96
Acetaldehyde 89
Acetals 101
Acetamide 111
Acetone 90
Acetyl chloride 108
Acetylene 25
Acetylsalicylic acid 115
Achiral 158
Acid anhydride 20, 107, 108
   formation 109
   hydrolysis 109
   naming 109
   reaction with alcohols 109
   reaction with amines 111
Acid/base extraction 83, 107, 127
Acid chloride 20, 107, 108
   formation 108
   hydrolysis 109
   naming 108
   reaction with alcohols 108
   reaction with amines 111
Acidity
   of alcohols 67
   of aldehydes and ketones 94
   of carboxylic acids 103–104
   of phenols 75
   of terminal alkynes 28
Acyl groups 104, 106
Addition reactions
   of alkenes 28
   of alkynes 28
Addition Polymerisation 34–35
Adrenalin 76
Adsorption chromatography 129
Alcohols 19, 58, 65–75
   addition to aldehydes 101
   addition to ketones 101
   bonding 66
   dehydration to alkenes 72–74
   formation 96, 113
   naming 65
   oxidation 68–70
   conversion to alkyl halides 71–72
Aldehydes 20, 33, 68, 89
   acidity of -methylene protons 94
   addition of alcohols 101
   addition of Grignard reagents 92
   addition of nitrogen nucleophiles 98, 99
   addition of oxygen nucleophiles 100–102
   aldol condensation 94
   bonding 91
   conversion to secondary alcohols 92
   derivatives 99
   formation from acetals 102
   naming 89
   nucleophilic addition 91
   protection 102
   reduction 96, 103
Aldol condensation 94–95
Alkaloids 89
Alkanes 2–4, 19
   bonding 5
   combustion 16
   conformational isomerism 7, 8
   constitutional isomerism 10, 11
   naming 3, 10
   halogenation 16
   reactions 16
Alkenes 2, 3, 19, 22
   addition of H-X 29
   addition reactions 28
   bonding 23
   chlorination 28
   formation from alcohols 72–74
   formation from alkyl halides 59–60
   hydrogenation 29
   naming 25
   oxidation with ozone 33

polymerisation 34
reactions 28
stereoisomerism 22, 23
Alkoxide ions 58, 67
Alkyl groups 10
Alkylammonium salt 58, 81, 84
Alkylating agent 58, 82
Alkyl halide 19, 56–64
  alkylation of amines 82
  elimination reactions 59–62
  formation from alcohols 71–72
  naming 56
  bonding 56
Alkynes 2, 3, 19, 22, 25–28, 58
  acidity 28
  addition of H-X 29
  addition reactions 28
  bonding 26
  chlorination 28
  hydrogenation 29
  naming 25
  reactions 28
  terminal 28
Alkynide ions 28
Alumina 129
Aluminium chloride 37–40
Amides 21, 107, 108
  hydrolysis 111
  formation 111
  naming 111
  natural 112
  primary 111
  secondary 111
  tertiary 111
Amines 20, 79
  acid/base extraction 127
  alkylation 82
  aromatic 86
  basicity 82
  bonding 79
  formation of amides 111
  formation of diazonium ions 86, 87
  formation from nitro compounds 86
  hydrogen bonding 81
  naming 80–81
  reaction with acid chlorides 111
  reaction with acid anhydrides 111
Amino acids 112
Ammonia
  reaction with acid chlorides 111
  reaction with acid anhydrides 111
Ammonium salt 58
Analysing filter 152
Analysis
  crystal structure 147–148
  elemental 130, 131
Aniline 36, 86
Anthracene 36, 44
Aromatic amines 86, 87
  formation of diazonium ions 86, 87
  formation from nitro compounds 86
Aromatic electrophilic substitution 37
  directing influence of substituents 42
  activating/deactivating groups 42
Aromatic hydrocarbons 19, 37, 112
Aryl halides 56
  from diazonium ions 86, 87
Asphalt 18
Aspirin 114
Basicity
  of amines 82
Benzene 36
  bonding 36, 38
  heat of hydrogenation 41
  oxidation of alkyl side chains 43, 45
Benzaldehyde 36
Benzamide 111
Benzoic acid 36
Benzonitrile 36
Benzoyl chloride 108
Bonding
  in alkanes 5
  in alkenes 23
  in alkynes 26
  in benzene 38
Bromonium ion (Br^+) 36–37, 39, 43
Butane 3, 4, 8, 9
Butanol 65
Butyraldehyde 90
Caffeine 89
Cahn-Ingold-Prelog convention 154
Camphor 104
Carbocations 30–31
Carbolic acid 76
Carbon dioxide 69, 107, 112, 113
Carbonyl group 89
  acidity of -methylene protons 94
  addition of Grignard reagents 92

addition of nucleophiles 91
bonding in 91
polarity 91
Carboxylate ion 104, 106, 107
Carboxylic acids 20, 103–114
acidity 107
acid/base extraction 127
derivatives 107
formation from alcohols 68, 112
formation from aldehydes 112
formation from aromatic compounds 112
formation from carbon dioxide 112
formation from acid derivatives 112
formation from Grignard reagents 112
naming 104, 106
reactions 105, 113
reaction with bases 107
reduction 113
Carotene 74
Catalytic reduction
alkenes and alkynes 29
deactivated catalysts 29
Cellulose 34
Chemical shift 144
Chiral centre 149, 150, 158
Chloronium ion (Cl$^+$) 37
Chromatography 129
column 130
gas (GC) 130
high pressure liquid (HPLC) 130
Chrysanthemic acid 115
Cinnamaldehyde 104
Cis-stereoisomer 16, 148
Citronellal 104
^{13}C NMR 141
Codeine 89
Combustion 16
Condensed structural formula 4, 12, 13, 14
Configuration 154, 156
Conformational isomers
anti conformer 7, 8
eclipsed conformer 7, 8
gauche conformer 7, 8
Conjugation 136
Constitutional isomers 10, 148
Crossed polarising filters 152
Crown ethers 79
Crystallisation 126
Crystal structure analysis 147

Crude oil 17, 18
Cyanide ion 58
Cycloalkanes 15
stereoisomerism in 16
Cyclobutane 15
Cyclohexane 15
Cyclohexanone 90
Cyclopropane 15
Daughter ion 132
DDT 64
δ scale 143, 144
Dehydration
of alcohols to alkenes 72–74
Derivatives of carboxylic acids 107–112
Dialkylammonium ions 82
Diastereoisomers 149, 160
Diazo dyes 89
Diazonium ions 86
conversion to phenols 87
conversion to aryl halides 86–88
coupling with phenoxides 88
Dienes 22
Diffraction 148
2,4–Dinitrophenylhydrazine 99
2,4–Dinitrophenylhydrazone 99
Directing influence of substituents 42
Distillation 126
Dopamine 76
Doublet (in NMR spectroscopy) 145
Dynemicin 27
Eicosatrienoic acid 114
Electromagnetic radiation 135
Electrophile 31
Electrophilic substitution 36
directing influence of substituents 42
activating/deactivating groups 42
Elemental analysis 130, 131
Elimination reactions 59–62
Empirical formula 130
Enantiomers 149, 153, 157, 158, 160
Enediyne antibiotic 27
Ephedrine 89
Ephinephrine 76
Ester 20
Ethane 4
Ethanol 65, 74, 92
Ethers 20, 58, 59, 68, 75
cleavage with hydriodic acid 78
properties and reactions 78

Ethoxide ion 68
Ethylene 22
Ethylene glycol 102
Ester 20, 107, 108
   formation 109
   hydrolysis 110
   naming 109
   reduction 113
E-stereoisomer 22
Extraction
   acid/base 127
Fermentation 74
Fingerprint region 140
Fluorouracil 64
^{19}F NMR 141
Formaldehyde 69, 90
   reaction with Grignard reagents 92
Formic acid 69
Fractionating column 126
Fragment ions 132
Free radicals 17
Functional groups 19
Gasoline 17, 18
Gas chromatography (GC) 130
Geminal diol 100
Geraniol 75
Glucose 101
Grignard reagents 62–63
   addition to carbonyl compounds 92
   reaction with acids 63
   reaction with aldehydes 93
   reaction with formaldehyde 92
   reaction with ketones 93
Halogen compounds 56–64
   naming 56
   bonding 56
Halothane 64
Hemiacetals 101
   reaction with alcohols 101
Heteroatom 56
High Pressure Liquid Chromatography (HPLC) 130
High resolution mass spectrum 133
Histamine 89
Homologous series 4
Hormones 21, 76
Hybridisation 5
Hydrates 100
Hydrazine 99

Hydrazone 99
Hydride ion 96
Hydrocarbons
   aromatic 2, 3
   bonding in 5
   branched 10
   hybridisation 5, 6
   linear 10
   saturated 2
   unsaturated 2
Hydrogenation
   of aldehydes and ketones 103
   of alkenes and alkynes 29
   of benzene 41
Hydrogen bonding
   in alcohols 66
   in amines 81
   in carboxylic acids 105
Hydrogencarbonate ion 107
Hydrogen NMR 141
Hydrolysis
   of acetals 102
   of acid chlorides 109
   of amides 111
   of anhydrides 109
   of carboxylic acid derivatives 107
   of esters 110
   of imines 98
   of proteins 112
Hydrophilic 66
Hydrophobic 66
Hydroxide ion 107
Hydroxylamine 98, 99
Identification of organic compounds 130–148
Imaginary mirror 150
Imines 98
Infrared (IR) 135
   absorption 139
   spectroscopy 139–140
Iron bromide 37–40
Isomerism
   cis/trans 16, 22, 148
   conformational 7, 8
   Diastereoisomers 149, 160
   E/Z 22
   Enantiomers 149, 153, 157, 158, 160
   Optical isomers 149
   stereoisomers 16, 148, 149, 159, 160

Isoprene 25
Isotope 133
IUPAC nomenclature 10
Ketones 20, 33, 68, 89
   acidity of -methylene protons 94
   addition of alcohols 101
   addition of Grignard reagents 92, 93
   addition of nitrogen nucleophiles 98, 99
   addition of oxygen nucleophiles 100–102
   aldol condensation 94
   bonding 91
   derivatives 99
   conversion to tertiary alcohols 94
   formation from acetals 102
   naming 89
   nucleophilic addition 91
   protection 102
   reduction 96, 103
Kekule 44
Kerosene 18
Limonene 25
Lithium reagents
   addition to carbonyl compounds 92
   lithium aluminium hydride 96, 113
Magnesium 63–64
Markovnikov's rule 30–32
Mass spectrometry 131–135
Mass to charge ratio 132
Menthol 75
Meso-isomer 160
Meta- 36
Methane 3, 4, 5, 16
Methanol 65
Methoxide ion 67
Molecular formula 133
Molecular ion 132
Molecular weight 131
Molecular orbital theory 28
Monomer 34
Morphine 89
Multiplicity (in NMR spectroscopy) 145
n + 1 rule in NMR 145
Newman Projection 7, 8
Naphthalene 36, 44
Natural abundance 133
Natural rubber 34
Nitration 37, 43, 86
Nitriles 21, 58, 59, 108
Nitrobenzene 36

Nitro compounds 86
   reduction to amines 86
Nitronium ion (NO_2+) 37, 43
Nitrous acid 86
NMR 141
Nuclear magnetic resonance (NMR) spectroscopy 141–147
Nucleophiles 31, 57, 58
   amines 82–84
Nucleophilic addition 91
Nucleophilic substitution 57, 58
Octane number 18
Olefins 2
Oil 17
Optical isomerism 149
Optically active compounds 152
Optical rotation 152–153
Ortho- 36
Oxidation
   of alcohols 68–71
   of aldehydes 102
   of alkenes 33–34
   of alkyl side chains 43, 45
Oximes 98
Ozone 33
Ozonide 33
Ozonolysis 33–34
Para- 36
Paraffins 18
Parent ion 132
Parts per million (ppm) 144
Petroleum 17, 18
Phenanthrene 37, 44
Phenols 36, 65, 75–76
   formation from diazonium ions 87
Phenoxide ions 75
   coupling with diazonium ions 88
Pheromone 21
Phosphorus trichloride 108
Pinene 25
^{31}P NMR 141
Polarimeter 152, 153
Polarised light 152
Polyacrylonitrile 35
Polymers 34–35
Polyethylene 35
Poly(methyl methacrylate) 35
Polystyrene 34–35
Polytetrafluoroethylene 34–35

Poly(vinyl chloride) 35
Primary alcohols 65
    oxidation 68–70
    from aldehydes 92, 96, 103
    from carboxylic acids 113
    from esters 113
Primary alkyl halides 56–57
Primary amines 80
Primary carbocation 31
Priority rules 154
Progesterone 104
Propane 3
Propanol 65
Propene 22
Propionaldehyde 90
Propoxide ion 68
Propyne 28
Propynide ions 28, 59
Prostaglandins 114
Protein 112
    hydrolysis 112
Pyrene 44
Pyrethrin 115
Quartet (in NMR spectroscopy) 145
Quintet (in NMR spectroscopy) 145
Quaternary ammonium ions 80–84
Racemic mixture 157
Radical chain reaction 17
Radiofrequency (Rf) 135, 141
R-configuration 156, 157
Reaction mechanism 30
Recrystallisation 126
Reduction
    of aldehydes and ketones 96, 103
    of alkenes and alkynes 29
    of carboxylic acids and esters 113
    of diazonium ions 87
    of nitro compounds 86
Resolution 161
Rf (radiofrequency) radiation 135, 141, 143
Rotation of polarised light 152
Sandmeyer reaction 86, 87
Salicylic acid 115
Salicin 115
Schiff's bases 98
S-configuration 156, 157
Secondary alcohols 65
    oxidation 68–70
    from ketones 92, 96, 103

Secondary alkyl halides 56–57
Secondary carbocation 30
Secondary amines 80, 83
Semicarbazide 98
Semicarbazones 98
Separation 126
Sequence rules 154
Silica 129
Singlet (in NMR spectroscopy) 145
Skeletal formula 12, 13, 14
Sodium borohydride 96
Specific rotation 152
Spectroscopy
    absorption 135–147
    infrared 139–141
    nuclear magnetic resonance (NMR) 141–147
    Ultraviolet-visible 136–138
Spectrum
    infrared (IR) 139
    mass 132
    nuclear magnetic resonance (NMR) 144
    ultraviolet-visible (UV-visible) 137
Splitting 145
Staggered conformation 7, 8
Starch 35
Stereochemistry 148–162
Stereogenic centre 150–155, 157, 159, 160, 161
Stereoisomers 16, 148, 149, 159, 160
    cis/trans 16, 22, 148
    Enantiomers 149, 153, 157, 158, 160
    E/Z 22
    Diastereoisomers 149, 160
Steric hindrance 7, 9
Steroids 21
Structural formula 4
Styrofoam 35
Sublimation 127
Substitution reactions 16
Sulfide 58
Superposable structures 150
Tartaric acid 159
Teflon 34, 35
Terephthalic acid 45
Terpenes 25, 75, 115
Tertiary alcohols 65
    from ketones 92
Tertiary alkyl halides 56–57

Tertiary amines 80, 83
Tertiary carbocation 30
Tetraalkylammonium salt 59, 83–84
Tetrafluoroethylene 34
Tetrahydrocannabinol 79
Tetramethylsilane (TMS) 144
Thiols 58
Thionyl chloride 71, 108
TMS 144
Tollens reagent 102
Toluene 36, 44
Trans-stereoisomer 16, 148
Trialkylammonium ion 83–84
Trienes 22
Triplet (in NMR spectroscopy) 145
Tyrosine 76
UV-visible 135
  spectroscopy 136–137
Valence bond theory 38
Velocity of light, c 135
Vitamin A 74
VSEPR theory 6
Wavelength of radiation 135
Wavenumbers 140
Williamson ether synthesis 76–77
X-ray 147
Xylene 44
Z-stereoisomer 22